ATOMIC ROMANCES,
MOLECULAR
DANCES

CHEMISTRY POETRY
by

MALA L. RADHAKRISHNAN

ILLUSTRATED *by* MARY O'REILLY

ISBN 978-1-4583-3192-2

The poems "Party of 5.55 × 10^1 mol/L," "Organic Matchmaker," "Ethanol, My Children," "PC Arrghh!," "Finding Amino," and "A Beautiful Amide" were reproduced from *Biochemistry and Molecular Biology Education (BAMBED)*, with permission of the copyright owner, IUBMB (the International Union of Biochemistry and Molecular Biology). Please see "Credits and Citations" for full citations.

For Samay

Remember, your life a state function is not;
'Tis the path that you follow that matters a lot.

Acknowledgments

For Their Starting Materials:

The students, faculty, and staff at Independence High School, MIT, and Wellesley College, Michael Altman, and Logan McCarty

For Their Catalysis:

Michael Brown, Ken Fastiggi, Kathryn and Galen Loving, Dale Shuger, and Wellesley College

For Their Element-ary Wisdom:

Chris Arumainayagam, Jim Davis, Robert Field, Vladimir Petrovic, Adam Steeves, and Judy Voet

For Their Kinetic Proofreading:

Katie Doherty, Jeff Langevin, Lucy Liu, Emma Nechamkin, Liz Parker, and Susan Yeh

For Their Strong Bonds of Love:

Sohil Parekh, Anand Radhakrishnan, Mom, and Dad

For Keeping an Ion the Illustrations:

Kyle Chiang

Credits and Citations

The following poems were reproduced from *Biochemistry and Molecular Biology Education (BAMBED)*, with permission of the copyright owner, IUBMB (the International Union of Biochemistry and Molecular Biology):

"Finding Amino." *BAMBED*. 33:2, p. 111, Mar. 2005. doi:10.1002/bmb.2005.494033022431

"PC Arrghh!" *BAMBED*. 33:3, p. 218, May 2005. doi:10.1002/bmb.2005.494033032433

"A Beautiful Amide." *BAMBED*. 33:4, p. 288, Jul. 2005. doi:10.1002/bmb.2005.49403304288

"Ethanol, My Children!" *BAMBED*. 33:5, p. 356, Sep. 2005. doi:10.1002/bmb.2005.49403305356

"Organic Matchmaker." *BAMBED*. 33:6, p. 419, Nov. 2005. doi:10.1002/bmb.2005.49403306419

"Party of 5.55×10^1 mol/L." *BAMBED*. 34:1, p. 34, Jan. 2006. doi:10.1002/bmb.2006.49403401034

The following poems previously appeared (excerpted/edited in the textbook and in full in the Teachers' Edition) in *Active Chemistry* (2007), It's About Time, Education Division of Herff Jones, Inc.:

"Moles Away From Equality," "Combust-a-Rhyme," "Sex and Acidity," "As the Magnetic Stir-Bar Turns," "No One Can Stopper," and "Days of Our Half-Lives."

The following poems were performed for a 2004 episode of the poetry video series P.L.A.C.E.S., produced by Philip Hasouris and james g. h. moore (some poem titles have been slightly modified to those shown here):

"The Ugly Doping," "Chemistry Never Dies," "Amalgam in the Middle," "The Flirt and the Inert," "Kelvin's Electric Guitar Concerto," "Party of 5.55×10^1 mol/L," "The Radioactive Dating Game," and "Guiding Light."

"Amalgam in the Middle" was originally published in *Technology Review* Magazine, Aug. 2005, p. M9.

"Reduced Hopes" previously appeared in *Somerville News*, "Lyrical Somerville," Mar. 29, 2006, p. 14.

"Days of Our Half-Lives" previously appeared in part or in full in the following publications:

Chemformation (MIT), Special Issue, Summer, 2004, p. 47.

CSGF Community, (Department of Energy Computational Science Graduate Fellowship Program), Vol. 1, Issue 1, Spring 2005, p. 3.

Lecture 32, Sylvia Ceyer and Catherine Drennan, 5.111 Principles of Chemical Science, Fall 2005. (Massachusetts Institute of Technology, MIT OpenCourseWare). http://ocw.mit.edu.

"The Radioactive Dating Game" previously appeared in the following publications:

Chemformation (MIT), Vol. 20, No. 1, Jan. 7, 2004, p. 2.

Tech Talk (MIT), Vol. 49, Issue 22, Mar. 30, 2005, p. 7.

New England Chapter of the Health Physics Society Newsletter, Vol. XXXXII, No. 2, Dec. 2005.

"Guiding Light" previously appeared in *India Abroad*, Oct. 28, 2005, p. M9.

Twenty of the poems here previously appeared in the chapbook *Chemistry for the Couch Potato*, copyright 2003, Mala L. Radhakrishnan and printed by Friends of Poetry. Some poem titles have changed since this appearance.

All poems have been slightly modified or edited since their previous appearances.

CONTENTS

**For the reader, I've labeled the poems that I deem
To contain somewhat bold or suggestive themes,
In case you are planning these poems to share
With the youngest corps of chemists out there.

6.022×10^{23}

I. The Mole and Stoichiometry

MOLES AWAY FROM EQUALITY

At the end of the elements' trendiest street
Was a beaker where just getting in was a feat.
The chances for entry were rather elusive;
The Club Atomic was very exclusive.

One day, ol' Tony, an atom of tin,
Went to Club Atomic to try to get in.
Along with his friends, he offered some dough
To the guard who snobbishly shook his head "no."

"Good-bye," said the guard with an air of odium,
A rather formidable atom of sodium.
"You're nothing but several unwanted pests;
You're not on the list of our pre-approved guests."

As he started to show them their shameful way out,
Tony intensely proceeded to shout,
"So what is this club's exclusive new rule
That gives you the right to be very cruel?"

"Well, clearly this club is a wonderful place.
We only have access to limited space,
So to stop overcrowding and bottleneck jams,
Per el'ment, we let in but only ten grams.

Ten grams of oxygen, ten of palladium,
The same for bismuth, iron, vanadium,
So on and so forth, you get the idea?
Now get outta here, okay? I'll see ya!"

"But wait just a moment," said Tony with care.
"Your admissions policy just isn't fair.
It gives fewer atoms the privilege to pass
When they have a higher atomic mass.

Now, compare us with helium (a mass of four),
And we have a mass of near thirty times more!
For thirty of them who get to have fun,
How many tins do you let in? Just one!

A much better way for you to be doin' it
Is using a different measure of unit.
Do *not* go by mass; instead you should try
To limit the *number* of atoms let by.

You *could* say 'per el'ment we'll let in a dozen,'
But you'll need to count on a much larger cousin:
The *mole*, a huge number—it's almost absurd:
Over six times ten to the twenty-third!

While a dozen of candles is twelve of that item,
A mole of candles … would take eons to light 'em!
Though it's still just a number and nothing else more,
Like a pair or a dozen, a trio, a score.

But why such a curiously odd quantity
Such as six times ten to the twenty-three?
Well, that is the a.m.u.'s it would take
For one gram of mass to equally make.

So one mole of helium, in grams, is four.
A mole of tin—well, thirty times more.
But you would be sure of ushering in
The same *number* of heliums as you would do tin."

While eyeing the book on his stately podium,
Thus said this sheepish atom of sodium,
"Fine, I guess we've got room now for you,
But I've got to kick out some heliums too."

So Tony and all of his friends were admitted.
From their lowly status were they now acquitted,
And all that it took was a little defiance
Grounded in fairness and knowledge of science.

LIMITING LOVE

Three moles of H_2 and two moles of O_2
Lived life in a bubble with nowhere to go to,
And what happens next, you might claim that you know:
$2H_2$ plus O_2 yields $2H_2O$.

Oxy tumbled around, was a victim of chance.
She was bouncing off walls in a gaseous dance.
Her "O" double bond had been ready for ages
To break so that she could then bond with some "H's."

14

She watched as H_2's were all passing her by
And wanting to react and couldn't, but why?
'Twas like the reaction was stuck there on hold—
Quite favored, but under kinetic control.

Though H_2 and O_2 did share that balloon,
'Twas like each was stuck in a little cocoon—
Two pitiful lovers who'd patiently wait,
'Cause they just couldn't form their transition state.[*]

But one day their home was attacked by a vandal
Who burst their big bubble by using a candle
That gave them the heat that they needed to start
The formation of water, fulfilling their hearts.

She excitedly danced as the gases got hotter.
"I'll finally bond with H_2 to make water!"
A big flash of light and a deafening boom,
And H_2O vapor flew all 'round the room!

As soon as the waters had celebrated
And thermally, all had equilibrated,
She hoped to be part of a nice H_2O,
But realized she hadn't reacted. Oh, no!

She had longed for so long and had been overzealous.
Of all of those waters was she now quite jealous.
"I thought I would follow the scripted equation,
So am I in chemical violation?"

From across the old bond did her mate start to speak.
"I know why you missed out on the goal you did seek.
To really be sure that you bring double-'H' in,
The O_2 must act as the *'limiting reagent.'*

15

See, we and H_2 must react with a ratio.
This ratio our balanced equation does say so;
The H_2 to O_2 must be two to one
For *all* to really experience the fun.

There were *three* moles of them and some *two* moles of us,
So 'twas not two to one—quite less, it was.
There were only enough H_2's, in fact,
For 1.5 moles of O_2 to react.

So H_2 was thus *limiting*, with us in excess.
Thus a half mole of O_2 did not get success.
In order to get all two moles to react,
There'd have to be *four* moles H_2, which we lacked!

So now you can see that there wasn't enough
For *all* the O_2 to react and form stuff.
'Twas just like that popular televised show
With a *bachelor* and twenty-five women or so.

See, all but one lost, while one of them won,
'Cause the mandated ratio was just one to one.
So he would be *limiting*, reacting away,
While the leftover women must wait for their day."

Thus, Oxy was sad, as she hadn't good luck.
For now, with another old "O" was she stuck.
But still there was hope of a match for her yet;
She needed to get on *The Bachelorette* !

'Note: This reaction is not necessarily concerted, and the
mechanism may involve multiple steps; there may be more than
one transition state.

AN AREA OF CONCERN

A mole of copper sat in a crucible.
They wanted partners who would be reducible.
But sadly, they'd see that only a fraction
Of them would experience a loving attraction.

When copper is heated, as many may know,
The copper and O_2 make CuO.
The love is quite strong and the friendship stable,
A match made across the periodic table.

The copper atoms had dreams every day
Of giving electronic dens'ty away
To an oxygen atom in O_2 gas,
A very electronegative lass.

But for most, the heavenly dreams ionic
Were crushed, 'cause often the truth is demonic.
Most of these atoms from loving were far,
As this mole took the form of a big copper bar,

And to get this reaction to go anywhere,
You need to have copper exposed to the air.
Two lovers can meet if each other they've sighted.
'Twas a party with only the *surface* invited.

So the surface turned a darker black,
But the rest, true love, they all did lack.

A frustrated atom from right in the middle,
Still single although he was hot as a griddle,
Thus spoke, "My friends, our woes can cease.
Our surface area we must increase!

We need a technician with implements manual,
With tools that can grind us all right into granules.
Our area'd be greater if we were a powder,
And Love, we could finally learn all about Her.

So let's rekindle our dreams ionic,
Relieve ourselves from this pain so chronic."
He summoned a mortar, a pestle, a file,
A student with gadgets to crush them with style.

An hour with several tools, it took her
To make this bar look like powdered sugar.
But she ground it so well, and the copper was heated
And mixed to ensure all the atoms succeeded.

The mass of the crucible's contents was higher.
The powder turned black atop the fire,
As each copper paired with an "O" as its bride
To form the copper (II) oxide.

But even this mighty reaction revealed
It's tough to near one hundred percent yield.
You can grind, can crush, can mash 'til you're weary,
But you'll merely *approach* the limit from theory.

So what of the copper who first made the case
For mashing the bar up in multiple ways—
The atom who desp'rately dreamed thoughts ionic?
He didn't react; perhaps that's ironic.

This plan's beloved mastermind
Was very sadly left behind.
So if *you* form a plan for a group to work toward,
Make sure you're the first one to reap the reward!

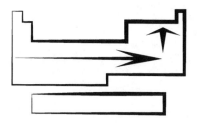

II. Periodic Trends

THE ATOMS' FAMILY

Let's take a look at my family portrait.
Up 'til today have I always ignored it.
My brother, my sister, my parents, and I
Are proud to be alkali metals. Here's why:

'Tis true that as atoms we're really quite large
By our meager effective nuclear charge.
But our size is a source of our family's pride;
Our electrons can wander both far and wide.

We're all rather ductile and malleable,
And this makes us really quite valuable,
Because we will readily ionize,
Becoming attractive, a positive prize.

There's sodium, my sister. She's over there.
Like me, she ought to be sealed up with care.
She'll react with water or even the air,
Giving her 'lectron up anywhere.

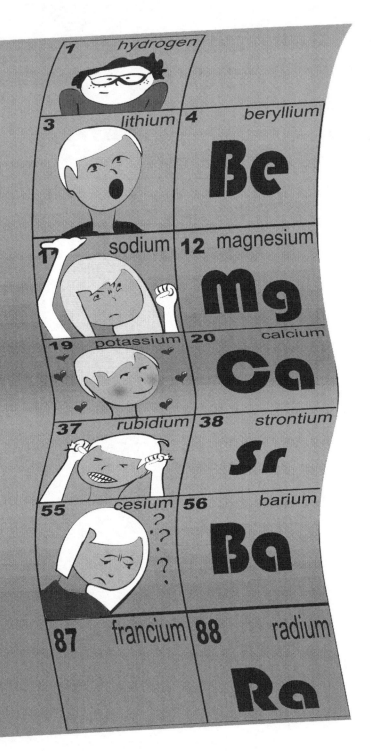

And here is potassium. He is my brother,
Ionically bonding with things like no other.
Quite social like I am, and much like my sister,
He just met a halogen, has already kissed her!

My parents, rubidium, cesium are they.
They're *dying* to give their electrons away.
As ours are their properties mostly the same,
But they're so extreme they can put us to shame.

You cannot see francium, here in this snap.
He left in a radioactive mishap.
The legend reports that he was unstable.
To me, his existence is merely a fable.

And there is a rumor that's having a run,
That hydrogen's my illegitimate son.
I swear he was left, though, on our front stoop,
And we're not really sure he belongs in our group.

Well, that is our photo, and here my tour ends.
You see, in my group there are patterns and trends.
The same likely holds true for you, you see—
"The apple can't fall very far from the tree!"

THE CHEMMY AWARDS

Gold: "Welcome, all, to the Chemmy Awards,
An event to which we always look forward.
Today is the day when we're finally able
To honor the best of the periodic table.

My name is gold, your host, you see.
(The trophies are made out of atoms like me.)
And now it is time to award the first prize,
Which goes to the atom of largest size.

Competing for being the biggest one:
Cesium, francium, radium … who won?
Which of the three worthy nominees
Will win tonight? … The envelope, please!

The winner is francium. Give him a hand!
But where is he, folks? I don't understand.
Well, francium is the most unstable
Of the first hundred elements of the table.

Because he's so rare, he's missing. Instead,
To accept the award for him, here's lead:"

Lead: "Thank you … francium worked to be large,
Through a small effective nuclear charge.
His valence electron he wanted to thank.
Its orbital helped him achieve this rank.

22

Atoms get big as you move down the chart,
As you add a new shell when each row starts.
But from left to right, the atoms contract
By incomplete shielding and more protons to attract.

Thus, the biggest atom resides
In the corner of the left, lower side.
The winner is francium—another it ain't.
And now I must go be in poisonous paint."

Gold: "Thank you, lead. ... and next up today
Is—who's interrupting? What's that you say?"

Cesium: "Yeah, this is cesium. I've been neglected.
Your faith in francium must be corrected.
You might suggest that he tackles me,
But some people think I am bigger than he.

Because he is much, much rarer than I,
This statement we can't really verify.
But y'all have some really big misconceptions.
With every trend, there will be exceptions."

Gold: "Sorry ... now, the next presentation:
The largest energy of ionization.
Or 'for which atom is it the hardest to free
The most loosely bound electron?' Let's see!

The winner is helium! She's absent, but why?
She doesn't react, so she must be quite shy.
To accept on behalf of the one who did win
Is our dearest old friend and a metal, tin."

Tin: "Helium deserves to win the prize;
This energy varies inversely with size.
If the 'lectron is to the nucleus near,
To be moved far away from the protons it fears.

So it makes good sense that the winner might
Be located way in the upper right.
This trend's exceptions are widely accepted:
Groups 2, 5, 8 are higher than expected.

A filled* subshell has the ability
To give an atom added stability.
This, along with an atom's size,
Tells you how readily it'll ionize.

I'm sorry, gold; I don't like to preach,
But the audience is reacting during my speech!
Their immodest dresses, exposed valence shells ...
The press takes pictures, 'cause this stuff sells.

And sodium and chlorine—creating a show,
Reacting right in the very front row!
As soon as these atoms think that they're famous,
They're suddenly feeling the need to be shameless.

Stop your reaction! I guess I should end.
But don't you forget your periodic trends.
They show that you needn't be so surprised
At how the table is organized!"

*(or half-filled)

III. Selected Elements

THE UGLY DOPING

Boris consistently felt like a fool,
Was teased at his own *element*-ary school.
The teacher would label him evil, depraved,
'Cause he simply couldn't be well-behaved:

"Your lack of respect, I just cannot fathom;
You ruin the name of all carbon atoms!
As punishment, you must go wait in 'time out.'
Now move it, and don't make my orbital shout!"

So every day he would go to that corner.
He constantly felt like an unwanted foreigner.
His classmates did look all exactly the same,
And to them, to tease him was just like a game.

"Hey, look at Boris," would start the abuse,
"He weighs just *eleven* a.m.u.'s!
But even with only five 'lectrons of charge,
He's still very unbelievably large!"

Those tinier carbon atoms would kid,
"You really destroy our crystalline grid!
Why don't you go on and find your way out?
You ruin our tetrahedral layout!"

Their jeers and derisions became systemic,
Insulting his talents academic.
"Boris is dumb as a vacant subshell,
And he can't even bond to things half as well!"

In school they were learning to form a diamond.
Their final exam was quickly arrivin'.
In order for them a nice diamond to form,
He'd need to follow the rules and the norms.

What they didn't know was that Boris would try,
But he'd fail to fit in and he didn't know why;
All others would bond to the carbons nearby,
But he'd stir them all up, making things go awry.

He'd carefully studied the teacher's directions
But hadn't the 'lectrons to make the connections
To help them create the consummate gem.
He simply felt … well, *different* from them.

Then, on the day of their final exam,
For which all the others did carelessly cram,
Boris, who'd studied, put forth all his might,
But he still failed to get a single thing right.

The teacher was saddened. "See, I had a dream
That my students may finally work as a team.
It's hopeless. I've given you all of the tools,
But your diamond breaks several textbook rules!"

And she pointed at Boris. "It's all your fault!"
Then somebody summoned her speech to a halt:
"My goodness, this diamond—it's gorgeous! It's only an
Arm's length away from display at Smithsonian!"

The teacher was baffled and offered corrections.
"I thought that this diamond had imperfections."
"Well, on a molecular level, that's true,
But it's due to those 'flaws' that this diamond is blue!

Diamonds of blue are rare indeed,
And the cause of the hue is a tiny seed
Of trace amounts of boron embedded.
This atom and others like him get the credit!"

He pointed to Boris, who eagerly called,
"I'm not a carbon, after all?"
And as his classmates cheered him on,
The "ugly carbon" became a swan.

"C" OF TRANQUILITY

Dear Diary,
Today I arrived at a startling conclusion,
Reflecting on things in a state of confusion:
See, I live a life very sheltered and closed,
With so much to which I am never exposed.

So boring has been my unwavering fate
As a carbon trapped in a carbonate,
Counting out every "tick" and each "tock,"
A sediment'ry lifestyle in limestone rock.

While other carbons go far and wide,
I've never left this mountain's side.
What different events that could have unfurled
Had I had a chance to see more of the world!

First, my friend Carrie, the sweetest of souls,
Spent part of her life in a piece of coal.
She'd keep on exclaiming that it was so "cool"
That she got to be part of a fossil fuel.

Then Carla … photosynthesis took her
Through many a ride to be part of a sugar.
She'd say, "I never realized
'Tis fun to be metabolized!"

And then there was Chlöe, the lucky doll,
Who got to be part of a "Bucky ball"—
C_{60}—indeed, to her surprise,
This cage helped win a Nobel Prize!

And Carl, he wandered off to a place
With others to make a nitrogenous base.
While speaking to me he would often say,
"There nothing like being in DNA!"

And Carlos, who says, "It's all about timin',"
Was lucky enough to be part of a diamond.
And Chris, whose life has been quite a ride,
Spent decades in poisonous cyanide.

Cat, who's been strangely missing since March
Is believed to be stuck in a store of starch,
While Candace—'tis even more puzzling than that—
Will likely be stuck for years in fat.

Candi spent time on the "dark side"
With an "O" in carbon monoxide.
Casey said he wasn't quite staffed right
When sent to be part of a pencil's graphite.

See, all of their lives have been packed with action,
Replete with exciting and novel reactions,
But me, my life goes not beyond
These stable, perpetual oxygen bonds.

But Diary, although I am feeling low,
I realize the beauty of "status quo."
When I look 'cross the bonds at my faithful "O's,"
Their comforting friendship to me really shows.

And I realize that *I* am the lucky one here,
To have had the same friends for some *billions* of years.
For *who* can claim such loyal brothers?
My story is just as complete as the others.

Indeed, when they told me about their reactions,
They said, "But those actions were only distractions.
It's time that we followed a stable route;
You're lucky, 'cause you've clearly figured it out!"

Diary,
The rest of the world I had wanted to see,
But now I see friends being jealous of *me*.
The lesson I finally do understand it:
Don't take the good things you've gotten for granted!

CHEMISTRY NEVER DIES

At the start of our tale is a suave atom who
Has now bonded with oxygen to make NO_2.
He floats in the air and pretends to pollute,
But really, he's spying on enemies to shoot.

And whom does our friend (known as "Bond") spy on?
As always, 'tis foreigners he has his eye on:
Germanium, francium, he sees them below,
And gratuitous violence proceeds in our show.

The enemy's disguised as if it's Halloween,
So Bond creates some trinitrotoluene.
He quickly connects with "C," "H," and "O,"
And the newly formed TNT screams out, "Let's go!"

They attack and create an explosive reaction,
The bad atom's fighters reduced to a fraction,
And "element 7" comes out with no scrapes,
Not knowing that one of the bad guys escapes.

But before our good story can further unfurl,
It's time for the entrance of Bond's special girl:
A diamondly atom of carbon, this lass
Is sparklingly shiny but sharper than glass.

The duo soon learns that germanium's alive,
And each wonders, "How did he ever survive?"
They now need to think of a plan that can't miss,
But given their passion, they soon start to kiss.

Their orbitals reaching all over each other,
They form a strong bond, then one more, then another.
An extra electron jumps on for the ride,
And with their octets, they've formed cyanide.

And so they lay low in their enemy's drink.
"Indeed, he'll be poisoned," at least so they think.
But hundreds of kilojoules at them are heaved,
And oddly, their strong triple bond has been cleaved!

"Where did she go?" but there's no time no look,
Because one of the bad guys got Bond on a hook.
'Tis sodium hydroxide, so vile, so base.
Its higher pH is now eating Bond's face!

But Bond quickly finds an "H" and three "O's,"
And the newly made nitric acid knows
That acid plus base makes water and salt,
And the neutralized enemy comes to a halt.

Now Bond needs but two things to win back his fame:
To kill germanium, then get back his dame.
So he bonds with an "O" who just happens to pass
And forms N_2O, or laughing gas.

Now germanium, taking his every breath,
Is slowly laughing himself on to death:
"Hahahaha! Bond thinks he will win.
I'll get him soon—" but he dies … with a grin.

Yet Bond can't enjoy the demise of his foe;
His missing woman is causing him woe.
He finds her in danger—it truly unsettles—
About to be killed by an army of metals!

"Don't worry, my dear. You're really all set.
They're alkali metals. We'll just get them wet!"
They quickly explode, forming base and H_2,
And the duo escapes to safety, on cue.

Now finally, Bond and his gal are alone,
But before they get close, he first binds to an "O."
You might ask why NO is needed, but trust me;
Just look up "VIAGRA" online, and you'll see.

As credits roll on, you walk out amazed
That Bond could fight all of his enemies unfazed.
Why do his victories so quickly amass?
Well, he got an "A" in his chemistry class!

AMALGAM IN THE MIDDLE

Silicon, sadly, was teased every day
In school, when atoms would line up to play:
Metals in one line, nons in the next,
But which should it join? All were perplexed.

Though like a metal (it was quite shiny),
Its conductivity was tiny.
Its band gap was too far from little,
And unlike metals, 'twas rather brittle.

It clutched electrons way too tightly,
So metals would tease it daily and nightly.
Yet nons would also jeer and nettle,
"You dress and look just like a metal!"

32

What pain 'cause it did not conform,
No box for it to check on forms.
Few others could know the lonely void
That it lived as a "metalloid."

But sili did not yet believe it was able
To socialize with the rest of the table.
Its valence shell did place it where
It had some *four* electrons to spare.

While greedy nonmetals weren't willing to share,
And the metals were willing to give anywhere,
Sili's electrons were things to be earned,
But they bonded with skill that just couldn't be learned.

So once other elements realized this fact,
Moles of them soon came around to react.
O_2 was the first to ask sili for dates,
And along with some others they made silicates.

The former outcast with hopes that were bust
Was soon the key in creating Earth's crust.
The pariah that had been given "the hand"
Was now in *every* grain of sand.

And nowadays, sili's still lionized.
Its band gap equals a perfect size
To dope with nearby brothers and sisters
And make computers from transistors.

As if its utility has not yet impressed us,
It's also in quartz and in glass and asbestos.
And silicon's used in chemical plants
For lubricants and breast implants.

34

Sili, its fourteen electrons so strong,
Proved all of its skeptical peers to be wrong
When it managed to move all the way out to Cali
And founded its very own aptly-named valley.

The "ugliest duckling" of the table,
Silicon basically couldn't be *labeled*.
So if *you* really think that you don't fit in,
Just think of silicon—*don't* give in!

THE FLIRT AND THE INERT

A bitter and jaded old atom indeed,
Crotchety krypton did not feel the need
To hook up or to be an electron donor,
For he had decided to just be a loner.

A miserly mind of the noblest rank,
His social calendar perpetually blank,
Complacently watching his grumpiness grow
As he sat at the end of the period's row.

One day, a seductive "knock-knock" at the door:
'Twas calcium, quite a notorious whore.
Her huge pair of 'lectrons was in his full view.
"I've pleased countless atoms," she said, "but not you."

Her 'lectrons were moving in crazy convulsions,
But krypton was feeling coulombic repulsions.
Said he, "I am one atom you'll never get;
I've *already* got what I want—an octet."

So she tried to induce his inebriation
By hitting him with some radiation,
To make him a couple of 'lectrons short
To see if he'd soon be the reactive sort.

But his valence electronic shell did prove
Too stable for even this whore to remove,
And as she was trying, the krypton would scoff,
"There ain't *no* one for whom I would take it all off!"

And with rage she yelled out, "I give up on this prude,
'Cause he thinks he's so great, but he's really just rude!"
And so calcium left, filled with rage in her core
(But she bonded with two of the bromines next door).

And krypton's abstinence seemed perpetual.
The other elements deemed him asexual.
Then one day, along came a couple, F_2.
"A guy with a guy?" thought krypton, "That's new!"

The fluorines together exuded such charm
That krypton, inspired, thought, "Hey, what's the harm?"
"You both look so happy. Can you give me some?"
And through covalence they formed a threesome.

This threesome most certainly broke town convention.
The octet police screamed, "You *need* intervention!"
But once they arrived, it was clearly too late,
'Cause krypton's valence was *ten* … and *no* longer *eight* !

But the KrF_2 floated blithely along,
And they strongly maintain that they did nothing wrong.
The rules of *chemistry* shape what's worth making,
While human-made rules are sometimes worth breaking!

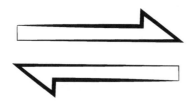

IV. Common Reactions

BRIDGE OVER TROUBLED H_2O

Sodium stood at the bridge's side,
Was contemplating suicide.
"I have such a pretty electron to give,
But nobody wants it, so why should I live?

My heart has for eons been painfully achin',
When first, I found out that dear chlorine was taken.
He quickly refused my electronic offer,
'Cause some other metal had filled his coffer.

Then fluorine came by, and we thought things could work,
But he left me for lithium, the horrible jerk!
And one by one, like an art form perfected,
The nonmetals left my electron rejected.

Too *much* for them was its coulombic caress,
The gem of my outermost subshell '3s,'
Which I'd doff like a gentleman's cap or his coat
For ionic romance of which *all* could take note.

But alas, my electron remained there, unseen.
I was forcibly put under kerosene,
And this unwanted abstinence—wholly unfair,
As I could not react even with the air!

A disgrace to my place on the periodic table,
To react, 'twas a fact, I was s'posed to be able.
My waning self-image now thoroughly killed,
'Cause I've not my chemical duties fulfilled.

How I'd sit, have a fit, all alone as I'd sob,
As I've been unable to finish my job;
With fluorine or bromine, with 'I' or 'Cl,'
My chances for love have been shot right to hell.

So I'm jumping in now; good riddance to strife.
Oh, may I be happy in afterlife."
And into the water she quickly did fall.
The result of this act, she thought would be small.

But after the jump, she was rather surprised;
This atom so hopeless had been oxidized!
Along with some others, electrons now gone,
She gave hers to H^+ to proudly don.

The reaction explosive, our atom now happy.
"Oh water, my love!" in a voice rather sappy.
The water replied, "I am free from the pain.
You have saved me from hazards of *acid rain*.

Before you jumped in, our pH was too low.
'Twas a bummer, no matter which way I would flow.
But then you came in, and you forced out H_2,
And now, this here lake is as neutral as new!"

Just reckon, she actually might have died!
But instead, she'd helped make hydroxide,
Had basically saved all the fish in the sea,
Had fallen in love, from her 'lectron now free.

Forever she'd swim in the arms of solvation,
In bliss with her happy new aqueous nation,
And she'd no longer be even slightly annoyed,
As her 'lectron was finally being enjoyed!

COMBUST-A-RHYME

Oliver learned all those "Don't's" and "Do's,"
The ones that get passed down through moles of O_2's,
As each generation was guided to trust
That its purpose in life was to make things combust.

The uppermost honor an O_2 could earn
Was to cause some organic matter to burn.
A kamikaze maneuver, this game,
With O_2 destroyed in creating the flame.

Oliver's brother reacted with methane.
His sister, well she earned her honors with ethane.
But which of the fuels would become *his* prize,
His prey, the one *he'd* oxidize?

He now was the last in his family remaining,
His self-esteem increasingly waning,
Was scared of being completely bereft,
Of being the *only* oxygen left.

One day, while diffusing around in the park,
He observed some equipment that started to spark.
'Twas no *matter* what caused this electrical prank;
It had gone and ignited a *propane* tank!

Now just when his prospects were starting to tire,
'Twas *his* turn to add some more fuel to the fire:
"I need to get close, to come into their view,
To help them make water and CO_2!"

But his motion was random, without a clear vision,
Was victim to all of the chance collisions.
Bouncing off friends, near elastic perfection,
He couldn't control his own motion's direction.

He shouted with vehement consternation,
"Oh, please take me toward the conflagration,
So I may receive the prestige so official
From doing my duty so sacrificial!"

And just as he finally got rather near,
On the brink of attaining his dream very dear,
The fire, ol' fate was to finish her;
Some guy brought out the extinguisher!

And Oliver realized that he had been benched,
Was blocked from entry, the fire now quenched.
The carbon dioxide, now right in his way,
Kept him from chemically hunting his prey.

Defeated, he plopped himself down on a surface,
Believing his life was devoid of all purpose,
But after awhile, he discovered a change:
He was stuck to the surface, and something seemed strange.

As Oliver checked out his changing environs,
He realized that he was now bonded to irons!
And part of a red-orange layer was he,
Unsightly Fe_2O_3.

That old iron bench onto which he did land
Was unusable now, and so people must stand.
And all that he wished was to make things combust,
But those dreams have been all covered over with rust.

SEX AND ACIDITY

She looked in the mirror and stared at her face.
It just wasn't easy being a base.
All that she wanted: a shoulder to cry on
And ways to remove her hydroxide ion.

Because of its stigma, her life was a mess,
'Cause never would she feel another's caress.
The guys she'd ask out would forever say no.
"I ain't goin' out with no bottle of Drano!"

Her molecular orbitals so unattractive,
Her hydroxide ion was not yet reactive.
All of her neighbors, they managed to hate her,
Except for her one friend, an indicator.

This friend would say, "It's tough knowing you.
Wherever we go, you keep me so blue.
But I've got a plan that can help you to score.
Let's both escape to the beaker next door."

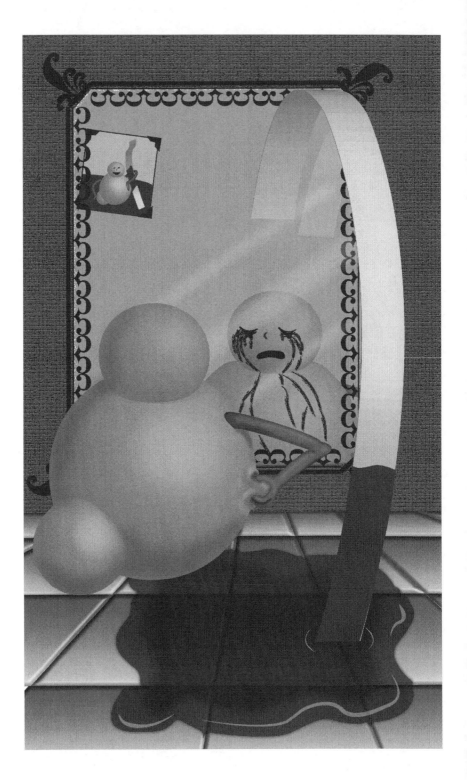

As they entered, a warning read, "Watch where you go—
The pH in this beaker is really quite low!"
And her friend was assailed by H^+ in the head,
And her color turned quickly from blue-ish to red.

The base didn't notice, 'cause she had her eye on
A very attractive hydronium ion,
And for the first time in her half-life, 'twas true
That a guy that she wanted now wanted her too!

But she moved really fast and could not wait to start,
And the next thing she knew she was pulled right apart,
'Cause the heat from the electrophilic attraction
Created a big exothermic reaction.

And the base soon arrived at the realization
That she was a victim of neutralization.
Forgetting their chemistry, two lovers at fault,
For now, they were nothing but water and salt.

Apparently, they were in clear violation
In causing the beaker's own annihilation.
As molecular souls thus drifted to heaven,
The pH returned to a boring old seven.

The moral is that you can get a lot more
By studying "chem" than by looking to score.
And I gotta admit, though I don't like to preach,
Chemistry's more pHun; that's spelled with a pH!

AS THE MAGNETIC STIR-BAR TURNS

A sodium atom walks into a bar,
His valence electron aspires to go far.
He's positive he will ionically bond
With a halogen partner of whom he is fond.

He winks at chlorine on the opposite side,
Who's sporting her outermost shell with pride.
Their mighty attraction they just cannot hide,
So they walk out a couple: sodium chloride.

The pair choose to go to a party that night
In water's big luxury beaker; a fight
Breaks out as the stir-bar revolves,
And the newlywed salt thus quickly dissolves.

The H_2O molecules break up the fight,
Surrounding the ions with all of their might.
As sodium sees the electron he lacks,
Chloride retorts, "You ain't *getting* it back."

Instead, she begins to flirt in the dark
With an ion of silver at the half-liter mark.
The two find so much they can chatter about
That they swiftly decide to precipitate out.

In sodium's nucleus, jealousy grows.
He wants silver chloride to decompose,
So he builds an electrolysis machine
To deliver the heartiest shock that they've seen.

When silver and chloride are deeply asleep,
Right into their bedroom does sodium creep,

And throughout their bodies the current does pass,
Producing some metal and free chlorine gas.

As sodium watches, his victory ensues.
"I turned that ol' chloride to Cl_2!"
But to his surprise he soon comes to find
That chlorine *likes* bonding with one of her kind.

The chlorines enjoy diatomic lives;
No need for a husband when there are two wives!
Though one day they suddenly get quite distracted,
'Cause to a magnesium they *both* feel attracted.

And as they *both* think his appearance a treat,
They wonder if they'll feel the need to compete.
But this Francophone atom to both says, "Bonsoir.
Have you ever heard of a *ménage à trois* ?"

So you realize it's not very much of a pickle
To see why the *human* heart is so fickle.
To understand this is no mystery,
'Cause what are we other than chemistry?

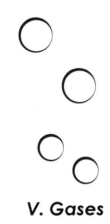

V. Gases

NO ONE CAN STOPPER

There's one thing my buddies and I can agree on,
That life's pretty good as an atom of neon;
No pining for love and no reason to flirt,
Because loners we are, so completely inert.

And my name is Ned, a magnificent neon,
Been bouncing about in this box for an eon.
The forces between us we barely can feel,
So a gas of our atoms is nearly ideal.

There's only one rule that we've clearly made known:
Don't mess with this gas—just leave us alone!
If you push us too far, you will certainly see
The power of "$PV=nRT$."

One day we were forced to impose this one rule
After (pardon the pun) we had "lost our cool."
I'll tell you the awfully painful story.
Beware, as the details are rather gory.

Long, long ago, we would peacefully roam
In a luxury test tube; she was our home.
'Twas roomy with crystal clear views, and atop her
Was placed a tightly-fit rubber stopper.

Now, life was serene; we had nothing to fear,
As our pressure inside was one atmosphere,
And the size of our home—few others could beat her—
A spacious hundred milliliters.

(Incidentally, we easily saw
That, thanks to the Ideal Gas Law,
'Twas true, at two hundred ninety-eight K,
0.004 moles of us lived there that day.)

But anyway, there we were, living our lives,
When a girl with a Bunsen burner arrived,
And the next thing we knew, its flame became blue.
We were heated so fast we didn't *know* what to do!

So we bounced all around with a newly found ardor.
Each time we hit glass, we now hit it much harder.
Our collisions with glass were all nearly elastic,
So the change in momentum was rather fantastic.

From a human's perspective, this translates to a measure
Of a really enormous increase in the pressure.
But you don't need a microscope's view to agree;
You just need "PV=nRT."

See, the heat invaded the test tube, and
Our gas wasn't given the right to expand.
The "T" was increased, and to keep the equation,
The "V" couldn't change, so the "P" saw inflation.

The pressure built up so severely, in fact,
That before the poor girl had some time to react,
CRACK! We pushed on the glass 'til it burst.
That day will live on in my mind as the worst.

By our beautiful home were we quickly bereaved
As the difference in pressure was swiftly relieved,
'Cause we quickly expanded, without an obstruction,
But as we flew out, we surveyed the destruction.

Shards of our home flew toward nearby lasses
Who weren't wearing their safety glasses!
The problem was thus exacerbated.
From all of our friends were we separated.

To imagine the graphic events makes me wince,
And alas, I've been quite bitter since.
The lesson, my friends, is a real no-brainer:
Don't ever heat a sealed container!

KELVIN'S ELECTRIC GUITAR CONCERTO

A mole of my helium fellows and I
Were minding our business, diffusing on by.
A frigid old bunch of friends were we,
Quite low in kinetic energy.

We relished each moment, taken slowly,
Every second sacred, holy,
Not rash, indulgent, or giddy, but thrifty,
Our temp in Kelvin a steady fifty.

And oh, how I loved the true peace in my life,
Bouncing so gently and free from all strife,
Gradually touring this box, our house,
To soothing sounds of Bach and Strauss.

At six hundred twenty-four joules per mole,
Our average energy proved we were cold,
And our "rms" speed, not a lively rate—
In meters per second, 'twas five fifty-eight.

Then one day we heard tawdry notes from afar,
The cacophonous scratch of electric guitar—
A rowdier group, unkempt for sure,
And boasting excessive temperature.

"You're here for us?" they heard me. "Yup.
We've come 'round here to stir things up,
A helium mole from across the tracks.
See, somebody thermally joined our flasks."

So there I was, now tossing and bouncing,
Victim to moshing, loudly announcing,
"This *all* must stop!" to which they would answer,
"Aw, what's the harm in some energy transfer?"

With every nearly elastic collision,
I hastened my pace with gymnastic precision.
Our group was quite hot and still heating up yet.
I would have been drenched, if atoms could sweat.

Than us was their energy 'leven times higher,
Though now they were visibly starting to tire.
As we flew 'round faster on average, they
Were slowing their pace in a serious way.

49

And soon did we reach our equilibration;
Our average energies were at their equation.
Our temperature now was quite simply the mean;
Five fifty and fifty 'twas halfway between.

'Twas three hundred Kelvin, the temp of a room.
For me, my new speed was indeed a fast zoom.
And they, they were forced to let go of their haste,
To take in just a hint of our cultural taste.

Though we also did learn from their frantic pace.
We realized that rock 'n' roll music has grace.
So together we started a new revolution,
An exciting, relaxing musical fusion.

We all did agree it was musical heaven,
Our rock 'n' roll version of Beethoven 7.
When opposites gather, new genres unfurl;
You learn and you pick from the best of both worlds.

VI. States of Matter and Intermolecular Forces

IN THE THERMAL ENERGY OF THE NIGHT

He looked to his left, and he looked to his right.
The scene never changed, from day to night.
His hydrogen arms, quite fixed in their places,
Were aimed at the same, old oxygen faces.

"I feel really trapped. It sure would be nice
If I weren't *stuck* in a block of ice.
I'd rather just end it all now with a pistol
Than live out the rest of my life in this crystal."

Several angstroms away in the lattice
Resided his secret love, named Gladys.
He wanted to bond with her and grow free
But was stuck in a world of low entropy.

Then one night, the sun just refused to go down,
And energy flowed to the crystalline town.
He started to wriggle around in his space,
And soon, he was flowing all over the place.

Now in a molecular mosh pit,
He hydrogen bonded with waters he hit
But hadn't control of his thermal motion,
And Gladys was sailing away in the ocean!

As neighbors around him quickly diffused,
He felt himself growing rather confused.
He found it a challenge to tell who was who,
And even his own self he no longer knew!

Gladys's oxygen nowhere in sight,
Each water's electrons were holding on tight,
But the sun ravaged through with more heat in its store,
And the waters were speeding their pace even more.

Some waters were flying away with a purpose.
Their energy overcame bonds at the surface.
The ultimate fate thus became a no-brainer;
Soon, they were bouncing throughout the container!

And thus, his location was clearly extinguished,
'Cause gaseous particles cannot be distinguished—
His pitiful soul, still present and trying
To be with his loved one and hold back his crying.

"So *this* is what it is like as a gas.
Perhaps I can find my Gladys, at last."
But as waters flew past, he thought, "This is bad.
I just can't tell Gladys apart from my dad!"

So he never was able to be with his lass—
Not as a solid, a liquid, or gas.
So if *you* are frustrated by love, well *tough*,
'Cause even on the atomic level, it's rough!

THE FOILING POINT OF WATER

Two liquids remained in their neighborly houses,
Their molecules randomly changing their spouses,
The left beaker filled up with pure H_2O;
In the house on the right did some methanol flow.

Ol' Wally lived life on the left side with pride,
A water who'd roll past his friends with a stride.
But as part of a liquid he'd look to the sky,
Because deep in his heart, he wanted to fly.

His soul mate who lived in the beaker at right,
Mark Methanol also thought often of flight.
"To fly? Oh, yes, 'twould be all the rage,"
Would say this CH_3OH.

Then one day, they noticed a change from the norm.
The beakers' inhabitants felt rather warm.
They started to speed up with motions frenetic,
The heat yielding energy, mostly kinetic.

And Wally screamed out with his "H's" agape,
"We're all speeding up, and I might now escape!
If there's enough heat added into this town,
The forces between us will not hold me down!

'Til then, I'm afraid I am stuck in this beaker.
It's too bad the forces between us aren't weaker.
To fly as a gas I must wait for so long,
'Cause our hydrogen bonds are so notably strong."

And Mark, who was thinking the very same thing,
Was hoping that freedom this new heat would bring.
But he, still grounded with motion too slow,
By intermolecular forces, was low.

Now just when the duo could no longer wait,
The dial was increased on their giant hot plate.
"How great! Perhaps we can now become free.
In degrees, it points right to sixty-five 'C'!"

The molecules speedily raced all around,
And some of Mark's friends lifted right off the ground,
Flying off in the air was this fast-moving faction,
Defeating their intermolecular attractions.

Still more of them left with excitement and cheer—
The vapor pressure: one atmosphere!
And Mark took his flight on that beautiful day
As the methanol rapidly boiled away.

But Wally looked on with an envious glare,
Still stuck in his beaker—it just wasn't fair.
While methanol floated, enjoying its luck,
At sixty-five Celsius, Wally was stuck.

So how come the methanol boils at this point,
But water can't seem to break out from this joint?
Well, water's got *two* "H's" bonded to "O,"
While methanol has only one, as you know.

So water can hydrogen bond like no other;
The waters so strongly attract one another.
The forces so huge in this beaker of water,
To boil you would need to make everything hotter.

So Wally's dream was thus shattered, deferred,
But then a most beautiful whisper he heard.
"Wally? My name is Wendy," she said.
"Flying you can't, but *bonding instead*?"

And together they flowed on as best as they could,
Just hydrogen bonding as waters should.
And though he did never lift off in a flight,
Once he found love, 'twas really all right!

LIQUID LONGING

My countless identical brothers and I
Do relish our vantage from way up high,
Each flying around as a nitrogen pair
And making up eighty percent of the air.

But sometimes it's boring, floating up here,
As part of the gaseous atmosphere,
Though really, I wouldn't complain, if only
It weren't so absolutely lonely.

Because we're a gas, it's clearly a fact
That our molecules rarely do interact.
Our density's measure is simply too low
For our intermolecular friendships to grow.

And if, perchance, two souls are to meet,
They scatter away without missing a beat—
No conversations, no talk of the weather,
No effort to blossom, to grow together.

But one day, however, I got the rare chance,
In the midst of our random molecular dance,
To talk with an H_2O just long enough
To be able to learn some incredible stuff.

He said he had buddies that he used to stick with,
While swimming in something he called a "liquid."
"In liquids, everyone's quite a bit closer.
You ever been part of one?" he asked. "No, sir."

He looked me over. "What *do* you expect?
You have no strong 'handles' with which to connect!
Hydrogen bonds you clearly won't make;
You lack polar 'H's,' for goodness sake!

If making a liquid were really your goal,
'Twould help if you sported some sort of dipole,
Or also, you could have been bigger in size,
So dispersion-type forces could dictate your ties.

But thus, these friendships that you seek
Can in your case, at best, be weak.
Your intermolecular forces, alas,
Are small. That's why you are a gas.

But wait, there's hope! The truth is that
It also depends what the *temperature's* at.
The higher the thermal energy,
The more will you fly all around and be free.

But when it's really, very cold,
Then weaker forces actually hold.
So go to a frigid locale, and then,
You will be liquid nitrogen."

And as rapidly as our exchange did commence,
He said, "Bye, I must go away now and condense."
But he taught me the cause of my gaseous deal:
My friendship conditions were just not ideal.

See, some can create many friendships with ease,
Without the need for help from a freeze.
For me, to get some friends clearly meant
I would need a different environment.

While many may dream of the sun and the heat,
I dream of the countless new friends I would meet,
How *I* would be popular, everyone's hero
At two hundred Celsius degrees below zero.

I do wish the change didn't need to be drastic.
To have friends right now would indeed be fantastic.
So if *you* are the type who can mingle with ease,
Just try to make friends with us too … please?

VII. Thermodynamics

THE (CHANGE IN) ENERGY WITHIN

I didn't have gloves, and I hadn't a hat,
Was stuck in a place with no thermostat.
I didn't even carry a coat, I fear.
My solution? ... Complaining, "It's *cold* in here!

I wish I could order the temperature higher,
Could turn on a heater or light up a fire.
Alas, there is nothing that I can do
As a helpless and frustrated NO_2."

Though I was the one who this sense was delivering,
All of the NO_2's in there were shivering!
"Our 'lectrons are freezing, if I may be frank; it
Would be so great if we just had a blanket!"

So there we all were, still going ballistic,
But one bouncing molecule stayed optimistic.
"I think I can help," said this soul very sanguine,
"With something I learned from this film about *penguins*."

I wished we'd had eyes we could roll at such words.
Was he gonna help us by mimicking *birds* ?
"I must have your nuclei really befuddled,
But trust me—what we need to do now is *huddle*.

The penguins, the film showed them making a form
That was tight and compact so they all could stay warm.
So everyone gather together and hug—
Like bundled-up babes, we'll stay toasty and snug."

We skeptically followed; each other we neared.
Then something transpired that seemed rather weird.
Our 'lectronic wave functions now rearranged,
As I noticed my nuclear places were changed.

I'd gotten a mate for the very first time; her
Voice was so pretty, "Alas, we're a dimer!
Your 'lectronic spirits will flourish and soar,
As together we prosper as N_2O_4!"

Then off together we skillfully glided.
With strangers we rather quickly collided,
And after a bunch of collisions so bold,
I noticed we no longer felt very cold.

So all of us loners, in love did we fall,
And a wave of warmth engulfed us all.
By pairing with mates who were all very sweet,
And then through colliding, we all made some heat!

My mate knew her chemistry; she was quite pleased.
"Energy, from forming our bond, was released."
Our molecular structure's now changed, you observed,
But the First Law says energy must be conserved.

So where could that energy presently lurk?
Well, it changes its form, through heat or work.
When all of us huddled, and pairs did meet,
A bit of our energy came out as heat!

The heat could be passed on to make us all warm
Through collisions within this gaseous swarm.
So now we are no longer cold anymore,
Thanks to chemistry, physics, and N_2O_4.

The crucial equation, if I may trouble you,
Is 'delta E equals q plus w.'
The use of this tool, it can be quite tricky;
The varying sign conventions are icky!

But once you define all your signs and your system,
The reaction's heat you'd predict with great wisdom,
So the notion that huddling would keep us from freezin'—
It worked, but for *thermochemical* reasons."

So my need for a jacket or coat thus averted,
Our goal reached with energy merely converted.
So if you've a challenge you're trying to win,
The energy needed, it all lies *within*.

I FOUGHT THE LAW (#2)

For whatever reason, I always do choose
To "do the right thing," as a goody-two-shoes.
But my angelic image I'd like to dispel,
'Cause *this* water molecule wants to rebel!

But what can this H_2O do to act out—
Poke holes in the beaker and plan to drop out?
No, I've got a scheme that will put 'em in awe;
See, I'm gonna break the *Second Law*!

Imagine: *I'm living my life unencumbered*
By fatalist Laws of Very Large Numbers;
I choose how my path is meant to be,
Am free from constraints of entropy!

So here I begin. I've managed to nab
The leader of this little chemistry lab.
I've drawn up a bartering deal to appease her.
In turn, she has put us all into the freezer!

And as the temperature slowly gets frigid,
Our overall structure grows to be rigid.
"Alas, my friends, now isn't it nice
To stand very ordered, frozen as ice?

And see how I'm snubbing the Law Number Two?
In microstates, ice numbers only a few,
So the chance that we'd randomly order and fall
Into place as a crystal is really quite small.

Our entropy now is quite clearly decreased.
I've broken the law, and I'm really quite pleased."
Indeed, I am proud of my act of defiance;
In one day, I've trumped many years of compliance.

But then says a hydrogen-bonding neighbor,
"I'd rather we not this point belabor,
But why would the scientists bother to make
A law that was *so* very easy to break?

Yes, ice is probabilistically worse,
Neglecting the rest of the universe,
But don't you forget, while H_2O freezes,
The entropy of its *surroundings* increases.

See, each little time that two waters do meet
To hydrogen bond, they're releasing some heat,
Which dissipates out to the crystalline borders
And to the surroundings, creating 'disorder.'

And heat will increase the 'delta S'
Of surroundings more when the temperature's less.
For our 'order' it overly compensates,
Making ice the likelier macrostate!

So at low enough temperatures, you'll see
That we will freeze spontaneously!
The system's and surrounding's entropies
Are captured in total by 'delta G.'

Alas, the chain of events that we saw
Thus *demonstrates* the Second Law.
So though you did try to control your fate,
The law did you basically *validate*!"

So now with this chemical lesson unveiled,
I realize that I have remarkably failed.
My only attempt to break a rule,
And I've proven myself to be a fool!

"But dear, to rebel, just listen to me,
And look up 'residual entropy.'"
And what happens next? Just wait and see,
In "I Fought the Law ... (Number Three)."

VIII. Solubility and Equilibrium

PARTY OF 5.55×10^1 mol/L

The H_2O house had a party one night
With three different compounds they chose to invite:
Hexane and sugar and salt were the three
Who were kindly requested to R.S.V.P.

And all on the guest list were pleasantly stunned
To be asked to the luxury beaker of fun.
They said, "Yes! Is alcohol permissible?"
"Of course," said the water, "for we are both miscible."

The sugars came first in a big silver spoon,
A bunch of sweet friends who had come just to swoon
Over H_2O guys who would make their spines tingle.
As soon as they got there, they split up to mingle.

Soon they were evenly spread all throughout
And surrounded by waters who'd holler and shout,
"You're so nice and polar; of you we're quite fond.
Let's have a good time as we hydrogen bond!"

Their "H's" and "O's" were suggestively close;
The entropy of the whole universe rose.
It's needless to say that amidst all the men,
The sugars didn't find each other again.

Next came the salt, which was in for a ride,
These married couples of sodium chloride.
The partying crowd made them question their mates,
And the salt began to dissociate.

The waters surrounded the split-apart ions
As if they were outfits divorcées could try on.
Their hydrogens pointed at negative chlorides,
While oxygens stood by the sodium's sides.

Now what was the cause of each sudden divorce?
Well, entropy served as the dominant force.
As time went along and the dating evolved,
Both sugar and salt were completely dissolved.

The last bunch to come was the hexane crowd.
They weren't expecting a party this loud.
The bonding and moshing and lust made them shout,
"Oh, *this* is not what chemistry's about!"

They huddled together to dodge the hysteria,
Minimizing external surface area.
The waters, too busy "H"-bonding to care:
"To give up these bonds for some oil's not fair."

So to their own kinds were such compounds so loyal.
Polar liked polar and didn't like oil.
The two didn't mix, but this we expect;
It's known as "the hydrophobic effect."

68

Sadly, society emulates "chem,"
As people keep mingling with others like them.
By breaking down boundaries this issue we'd solve,
And most of our problems would simply dissolve.

UNDISSOLVED MYSTERIES

I've been such an angry old ion of barium
Since two little fluorines asked me to marry 'em.
Each asked for an electron from my "6s,"
And as we could all have octets, I said, "Yes."

So now in a crystal of BaF_2,
I'm ionically bonded—there's nothing to do!
So bored in this grid, I want to be free.
I regret this decision's effects upon me.

When I was a younger and happier barium,
I used to go swim in the local aquarium
But was ousted upon an examination,
Was labeled "hard water contamination."

I miss being wrapped up in H_2O,
But here I am now, with nowhere to go.
If *you* were in my place right now, you'd all hate it.
Oh, how I long to be freely solvated!

But wait—do my senses deceive me? A beaker!
I knew luck would help out the needy who seek her.
A liter of water—it's filled to the brim;
I might be able, again, to swim!

69

To be sure of my prospects for happy solvation,
I've managed a chemical calculation:
Two grams of crystal—that's 0.01 moles,
Less BaF_2 than this water can hold.

See, I know our compound's K_{sp},*
So if 'x' equals *solubility*,
My chemical powers can help me declare,
"Our K_{sp}'s x times $4x^2$."

So x is 0.018 moles of us.
We're less, so we should dissolve with no fuss.
Okay, my crystal, 'tis time to start
The dissolving process of splitting apart.

We're now in the water, but something is wrong.
I've been here awhile now, been waiting so long,
But only a fraction of us is dissolved.
This mystery I must now attempt to resolve.

Ah, here is a water that I will now beckon:
"Hey, water—yeah, you—come here for a second.
I currently want to investigate
Why I remain in the crystalline state.

Why haven't we all become aqueous?
The math says you would take me, yes?"
The water replies, "You would be correct,
Were it not for the *common ion effect*.

You simply assumed we were pure H_2O.
Did you even ask once to make sure? Oh, no!
Well, there're already ions dissolved in here:
0.5 moles of sodium fluoride, my dear.

Now, usually, the presence of others wouldn't matter,
But your group has fluoride and it also had her.
You've ions in common; you'll need to redo
The math you had done and start anew."

"Oh ..." so out comes my calculator.
Darn that fluoride, how much I hate her!
Her molarity's high, and our K_{sp}
Demands, therefore, that there's less of me free.

x times (0.5 plus $2x$) to the two
Should equal the same K_{sp}, 'tis true.
But I learn from this newer equation I solve,
That less than 0.018 *grams* can dissolve!

Well that now explains this mystery,
And I guess that I've learned some chemistry.
This "common ion," it makes me upset.
Now all of these actions I've come to regret.

I might have my freedom occasionally, though
(An equilibrium's dynamic, you know).
But nevertheless, I'll mostly be stuck
As a solid. Oh well, so much for luck.

So here's some advice you should listen to.
Make sure you all know what you're getting into.
When compounds compete, it'll never seem fair,
'Cause as people know, it's hard to share.

---

*The K_{sp} of barium fluoride at 25°C is 2.4×10^{-5}, according to Zumdahl, S. *Chemical Principles, 5th Edition*. Boston: Houghton Mifflin, 2005.

OH, K!

As a sulfur dioxide molecule, I knew
I was able to bond with a nice Cl_2.
But the K for this marriage was sadly no more,
At six hundred Kelvin, than 0.164.

So one day my friends and I jumped in a flask
With some Cl_2 gas, and we realized our task:
To bounce all around and absorb all the fun
Of allowing two wave functions to bond into one.

To several partners did I say, "I do."
And each time we formed SO_2Cl_2.
But soon we'd predictably feel that old "force,"
When entropy'd randomly cause our divorce.

And soon the rate of the wedding attending
Would equal the rate of the marriages ending.
The forward reaction's rate was no worse
Or no mightier than the reaction's reverse.

The two rates did vary until their equation,
At which point we had reached equilibration—
When the quotient of couples by the product of unpaired
Was constant at 0.164, we declared.

But though the *amounts* of the compounds stayed fixed,
From sep'rate to married we still moved betwixt.
So it wasn't like I could just "chill" in a hammock—
A chemical equilibrium's *dynamic*.

But being in equilibrium's nice,
And all is so peaceful—no need to think twice.
So you can see why I was rather disturbed
When one afternoon, our state was perturbed.

More Cl_2 gas had been added that day,
And therefore, our quotient didn't equal our K.
With some did we bond; they were lovely and gifted,
And toward the product the reaction thus shifted.

Then someone ejected some SO_2 friends;
To bring back the K, did some marriages end.
At making our quotient be K were we deft;
To reach it our reaction did shift to the left.

But oh, 'twas not over, as you all might think.
The size of our flask, it started to shrink!
And whenever a volume decreases its measure,
There is (from Boyle's Law) an increase in the pressure.

So what could we do to oppose this new change?
Reacting at such a high pressure was strange.
'Twas a snap to collide so a marriage was fast,
And we shifted to products, made fewer moles of gas.

Phew! Again, we settled to K,
But just when I thought that this peace was to stay,
A scientist started to give us some heat,
And enthalpy, entropy, both did compete.

This heat clearly ruined the balance of rates,
So would *more* of us pair up with Cl_2 mates?
The creation of order was harder that day—
A new temperature changes the value of K.

And it turned out that couples broke up to make hermits,
Because this reaction was exothermic.
The value of K, indeed, was decreased,
And a net count of marriages actually ceased.

So now you can see why I'm always so manic;
I live at the whim of *statistical mechanics*.
Oh, how I hanker so much for the day
When my life won't be ruled by Le Châtelier!

IX. Electrochemistry

CHEMFOMERCIAL

Have you ever been stuck in the very last row
At a crowded, chaotic, and costly show?
You struggle to see; you can't hear so well.
Well, that's how we felt in our valence shell.

My pal and I know that you can't really fathom
How boring it is at the back of an atom,
But trust me, though protons may positive be,
Their attraction from way back is real hard to see.

Within a magnesium atom we stayed,
But the twelve shielded protons were too far away.
It's rough at the end, you see, it's true,
Of $1s^2 2s^2 2p^6 3s^2$.

As we knew we'd remain there in really bad shape,
We thought it was high time for us to escape.
So from the back row came our program of action:
We planned to create a *redox* reaction.

But we lacked all the chemical information.
We knew that we needed an education,
So we called "555-C-H-E-M."
They helped us to learn and to master our "chem."

A newspaper ad did they help us produce:
"We're looking for partners who will be reduced."
Within a few hours an adoring lass said,
"Hi, I'm hydrochloric acid!"

We pondered her offer and finally kept her,
Her proton the perfect electron acceptor.
We left the "Mg," and its cloud was now sized
Rather small because it had been oxidized.

But because there were two of us looking to part,
And each proton could only hold one near its heart,
It took two HCl's for the oxidation.
This ratio thus led to a balanced equation.

Now after the 'lectronic transfer was done,
The newly made hydrogen atoms made one
Single unit of H_2, which managed to hop
In a bubble with others to float to the top.

But what had become of the atom we parted?
'Twas quite a bit different from whence this poem started.
The magnesium ion, dissolved, was like new,
In aqueous $MgCl_2$.

So that's what we did, and as we have shown,
We both now have nuclei that we call our own.
No more of the shielding, no need to compete,
As we *each* have our very own front row seat!

So if *you* are upset by your current location,
It's time to consider a different vocation.
Try "chem"—you will see, as we showed in this stunt,
That chemistry helps you to come out in front!

Commercial Disclaimer:
(An atom's electrons are identical, in fact,
So to speak about "this" one or "that" one is whacked,
'Cause every electron in "Mg" is the same.
For implying it's otherwise, I am to blame.

But still, the "3s" subshell is full,
And even though naming who fills it is bull,
It's useful to see things from "their" point of view,
But don't take this analogy too far through.)

CHARGED WITH A SALT AND BATTERY

Life in a battery's tough, it's a fact.
I have all this pent up desire to react.
But I must admire my love from afar
And wait until somebody starts up this car.

My oxidation state's plus four.
I desp'rately want two electrons more!
And then comes a voice: "My PbO_2,
I've got two electrons—they're just for you!"

But I'm at the cathode, and he's over there
At the anode—I tell you it's really not fair
How humans, they force us apart every hour,
Exploiting our smitten ol' hearts to get power.

I'm stuck here, surrounded by sulfuric acid
And forced to obey and remain nice and tacit.
Humans are heartless; they're all nuts and bolts.
To them, I'm worth 1.685 volts.

And metallic "Pb," my love so prized,
He really just wants to be oxidized.
Just like my woe, his pain has a label—
0.356 volts—to make himself stable.

In adding together our half reactions,
You get about two volts of car-driving action.
And that ain't enough, as those humans do know
They get *twelve* volts with six longing cells in a row.

We all want to create some lead (II) sulfate,
And we at the cathode would normally wait
'Til our owner decided to go out somewhere,
And then, we could get our electron pairs.

But I am not patient; I want to get even,
But it could be days 'til the owner is leavin',
So I've got a plan to get "e's" from my lead—
'Twill render the owner's battery dead.

I really do hope that my scheme will work; it
Involves creating a big short circuit.
And then, the current will come real fast,
And I can obtain his electrons at last.

(She connects the two ends and prepares for the fun.
Within a short while she regrets what she's done!)

Yikes, I think I should probably pray,
'Cause too many 'lectrons are coming my way.
They're putting out heat as they come down the road.
We're starting to melt, and quite soon we'll explode!

Oh, man, I totally feel like a fool.
I should have remembered my physics from school.
A wire's resistance is almost not,
So currents get huge, and it all gets real hot.

The 'lectrons are flying this way like a blur.
Two 'lectrons I've nabbed but know not whose they were!
So gross—they could be from *any* old boy;
My dearest love's pair, I'll never enjoy!

So that's what I get now for being too greedy,
But humans, it's *your* fault for making me needy.
My love and I now will be always apart,
But at least the owner's car won't start!

REDUCED HOPES

I think of her name, and I instantly miss
Her attractive and vibrant electronic kiss.
The bond was intense, and nothing could stop her—
At least, so thought *this* atom of copper.

We had been married with compounded glee.
Then one day, she stated, "I need to be free;
If we stay together, my future seems bleaker,
So I've thus decided to jump in The Beaker."

From deep in my core came a passionate frown,
"The Beaker? *That singles bar downtown?*"
But I couldn't stop her uncomfortable grin,
So I did what I could, and I followed her in.

I couldn't believe her; was this all a joke?
As soon as we entered, our bond quickly broke,
And away she diffused, now sep'rate from me,
Searching for 'lectronic love in the sea.

In transit, she'd bump into guys here and there.
They flaunted their charges, pretending to care.
The thought of her elsewhere, how deeply it haunted,
As I'd been so sure it was *I* that she wanted.

But I knew as long as those singles may dance,
If I remained positive, I'd still have the chance
To receive my love back to the desire so distal
Of someday creating our very own crystal.

But off from the flower did fall the last petal,
When who came along but aluminum metal:
'Twas lighter, brighter, in fact, more active;
Its presence would render me quite unattractive.

My positive outlook now overturned,
As the negative energy quickly returned,
And I, a wallflower, almost inert,
Was too atomically ugly to flirt.

I watched my ex-wife have a torrid affair
As I joined other coppers to form a thin layer:
The rejects, the losers, the ones they ignore,
Unlike fish in the sea, we were all cast ashore.

And meanwhile, the water was boiling away,
And the singles paired up in an orderly way.
As water discreetly evaporated,
The crystalline families were thereby created.

But I remained longing, without a reaction,
Aloof from the happily bonding faction.
My love now in bonds so compromised,
While I was still chemically paralyzed.

I never experienced her forces again,
So now I must dream of the time way back when,
When she and I first commenced with our dating,
Before all this tragic *copper plating*.

And oh, how I reach an excited state,
Recalling our mutual eigenstate,
But now I suppose I will just have to settle
For living as lonely, copper metal. ...

X. Kinetics

L'AMOUR, THE BARRIER!

William the Water was always too shy
To go up to an oxygen just to say, "Hi.
I'd *really* love to react with you.
Together, we'd cook up some H_2O_2."

He'd watch them diffusing around every day,
While thinking of clever one-liners to say,
Like, "Ain't no molecule in this wide ocean
Who struts with your sexy Brownian motion."

But just as he'd see any oxygen near,
He'd quickly withdraw himself, taken with fear.
Approaching an oxygen—nothing was scarier;
His shyness, it caused a high energy barrier.

As humans, how do we lower this wall?
Well, sometimes we guzzle some alcohol.
It acts like a catalyst, freeing that gate,
To stop inhibitions and speed up the rate.

But Willy's rate could not have been slower.
He needed to make his barrier lower.

But everyone knows that catalysts act
To lower the barrier so things can react.

So he went to the market and searched low and high
For catalysts claiming they'd make him less shy.
He finally found that the iodide ion
Was clearly a catalyst he could rely on.

He sprinkled some into his hydrated home.
The next thing he knew he was reading a poem
At oxygen's side—the catalyst worked!
But oddly enough, she appeared rather irked.

"Bonding with *you* ? I will not be able.
Hydrogen peroxide's grossly unstable!
I'd rather be stripped to a charge of plus two
Than to have to endure a reaction with you."*

William was totally taken aback.
The catalyst should have installed him on track.
So why did this scheme end up making him panic?
Well, he forgot about thermodynamics.

A catalyst functions to speed up the rate
Of revealing a substance's ultimate fate.
But that ultimate fate is perpetually sealed,
And to no single catalyst could it ever yield.

So if a reaction is fated to be,
A catalyst causes its speed-up, you see.
But some are not fated to happen at all;
That's how catalysts differ from alcohol.

*This is an exaggeration, like "I'd rather die than be with you."
Second ionization energies are extremely high in general.

SPEED DATING GETS A "D"

I found it so hard being all on my own,
Diffusing around the block, alone.
My ex used to say, "I'd marry her!"
But was stopped by an energy barrier.

So one day I actually found myself waiting
In line for none other than, yes, "speed dating."
I figured I might as well give it a try,
By turning the magnetic stir-bar on "high."

The place grew crowded as in we diffused,
Filled right to the brim with lonely H_2's.
My nuclei nervously spun in delirium,
For I was the only present deuterium.

Competing for romance, I realized my fate.
The partners would think I was quite overweight.
So rare was my kind, amidst this zoo,
As only my "H's" had masses of two

(And that was measured in a.m.u.).

Now in came the partners, from far and wide,
Diffusing about with molecular pride.
I needed to feel more uninhibited;
My time with each partner was very limited.

I soon felt there must have been something I lacked.
The partners with others did quickly react.
Their urge was so strong; though they'd all try to fight it,
They still ended up being very excited.

The bond in H_2 is so squirmy and wriggly,
Vibrating backward and forward so quickly.
No wonder they found it so easy to break
And react with the partners to novel things make.

But I was, because of my heavier mass,
A kinetically paralyzed little lass.
I wanted reaction, to savor the pleasure,
But *just* couldn't handle the speed-dating pressure.

My atoms' vibrations were simply too slow
To speedily get a reaction to go.
So I left the event without a new mate,
While each of the others went off with a date.

Sad and alone, I started to mope,
Cursing the mass of my isotope.
(In truth though, because our abundance is low,
We're actually worth a *lot* more dough!)

But while I was moping, along came a lad
With pouting electrons who said, "I'm sad.
I tried 'speed dating,' and *man* was it rough;
I just wasn't able to choose fast enough!"

I studied his orbitals with a keen eye
And realized he might be the *perfect* guy!
And somehow, both of us seemed to know
That we needed to take things nice and slow.

We took our own time, indulged our will,
Waited to cross the "kinetic hill."
And thermodynamically, *boy* did I score,
'Cause good things are surely worth waiting for!

ORGANIC MATCHMAKER

I'd really given it all that I had.
I flirted at bars, wrote a personal ad.
But every molecule I would encounter
Seemed wrong for me once I had learned all about her.

While some would appear quite attractive and able,
Upon reacting we'd end up unstable.
Every time a reaction would start,
Ol' thermodynamics would push us apart.

I knew, though, that there was a fish in that sea,
Was someone who'd surely react with me,
Proceeding forward spontaneously
With highly negative "delta G."

But finding that someone? My odds were pathetic.
The issue I faced here was mainly kinetic.
Imagine two lovers apart, in seclusion
And forced to meet only by random diffusion;

Yet even that pair wouldn't have it as rough.
In my case, just meeting was still not enough,
'Cause we had additional obligations—
Colliding with adequate orientations.

And oh, how I longed for romantic involvement,
So lonely, surrounded by aqueous solvent.
Then one day—'twas no way that I could mistake her—
Along came my magical "matchmaker."

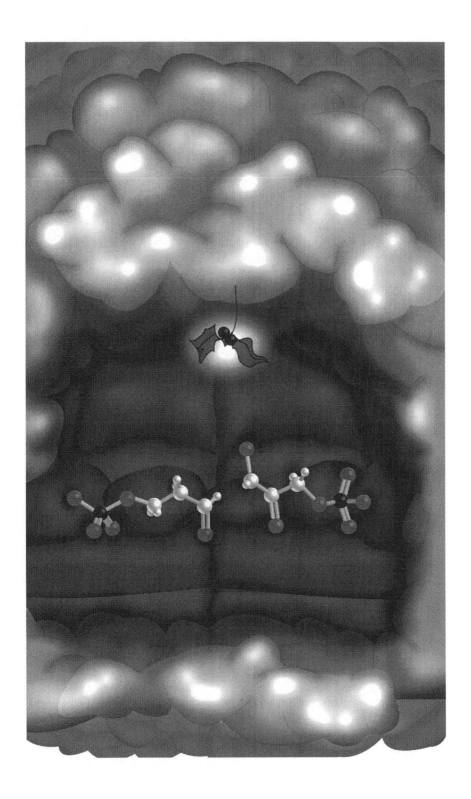

A towering enzyme whose active site
Could welcome two molecules nice and tight
Like a *loveseat*; we'd perfectly fit, in fact,
Positioned ideally to quickly react.

The enzyme was meant just for me and my match.
To her both my partner and I could attach,
And she helped me react with my ultimate mate
By stabilizing our transition state.

She stated, "My job is the best occupation;
Reducing your energy of activation."
And then she said, "Although it's been fun,
My job as your catalyst here has been done."

So if *you'd* really like to react with a purpose,
Just try the "enzyme dating service."
'Tis true that I found mine by accident;
She came, she helped, and back she went.

But maybe an enzyme can help you out too,
Can help introduce you to someone new,
Creating a setting that's fit for a kiss
And resulting in thermodynamic bliss!

XI. Subatomic Particles and Nuclear Chemistry

THE ION WITHOUT A NAME

Sadly, she wandered the town without aim,
For she was an ion without a name,
A vagrant for whom none would put on a fuss;
When asked who she was, "I'm 'Anonymous.'"

A couple of times she had guessed and been wrong
'Bout the square on the table to which she belonged.
"Cr"? "Ba"? "N"? ... or "V"?
Her nucleus *ached* for identity.

One day, an atom she met on a bus
Said, "What is your name?" "I'm 'Anonymous.'"
"My dear, what a shame, such a beautiful dame;
I'll help you try to discover your name."

He counted her 'lectrons aloud for the kicks,
"$1s^2$, $2s^2$, $2p^6$ ...
$3s^2$, $3p^6$, hmmm ... eighteen.
Alas, you're argon, the noble queen."

91

He left her abruptly, recalling the tales
That noble gas hearts are all harder than nails,
And she was let down, her hopes far gone.
"Of all of the choices, oh, why *argon* ?"

A curious atom sat down by her side.
"My dearest, you look as if someone's just died!
What is your name?" Her voice sounded ominous.
Our ion whimpered, replying, "'Anonymous.'

They wanted to label me 'argon,' 'tis true."
To this she replied, "Oh, no! Not *you*.
I bet you they counted electrons; if so,
They're wrong, because those often come and go.

The number of 'lectrons you cannot rely on
For figuring out the right name for an ion.
But I've an idea, 'twill all surmount.
Instead, we'll choose your neutrons to count.

Alas, it appears that you've got twenty;
Indeed, you're the *witch* from the stories a-plenty.
The legend says 'gone are the days so halcyon
Once one crosses paths with the evil calcium.'"

Panicked, the woman then got off the bus,
Our ion remaining, her name now a cuss.
But still in her ground state, she tried to be strong,
Was hoping the judgmental woman was wrong.

And next came a man with a monstrous cloud,
A zoo of electrons that cloaked like a shroud,
A subnanoscale hippopotamus.
"Your name?" he questioned. "'Anonymous.'

I heard I was calcium ... another mistake?
She counted my neutrons, for goodness' sake!"
"My dearest, today you're the luckiest lass—
The neutron count merely affects your *mass*.

The number of neutrons has nothing to do
With the actual answer to 'Who are you?'
'Cause atoms who share the very same name
Need *not* have their neutron amounts be the same.

To deal with this maddening issue, one copes
By using the cryptic term 'isotopes,'
Or, atoms who live at the very same square,
Although their masses they simply don't share.

But trust me, 'cause I've got the optimal way
To determine your actual name, today.
It may be quite hard with your 'lectronic coat on,
But I'm going to venture to count every proton.

The name of an atom will only depend
On the number of *protons*. Period. The end.
Regardless of 'lectrons or neutrons you own,
The proton count renders your name in stone.

And you've got nineteen of them ... beautiful day!
A potassium ion—attractive, I say!
And I, an iodide ion," he stated.
(A bond quite ionic was thereby created.)

She stared at him proudly, happy and googly-eyed.
"A bond will forever connect both our nuclei!
And love with its magical presence shall honor us,
And I'll never, ever, again be 'Anonymous!'"

MY RADIOACTIVE BROTHER

My brother and I are two atoms fraternal
Who used to believe that our youth was eternal.
Our childhood excursions were always carefree
As we would diffuse in a vacuous sea,

Immune to worldly intoxication
And rarely experiencing oxidation,
Were under each other's constant surveillance
As if we were bonded through covalence.

Now, as a potassium atom I know,
For us, this covalence cannot be so.
But nevertheless, we were best of friends,
Believing that innocence never ends.

Now sometimes, I'd notice that my dear brother
Was just a bit slower than me and the others,
On average, arriving a little bit late,
For he was stuck carrying extra weight.

See, of the potassiums that we both knew,
Only *he* had an extra a.m.u.
Some of our peers believed he was dubious
By the extra neutron within his nucleus.

At first, things were normal, more or less.
But then my dear brother got really depressed.
He felt like his life was a sham, you see.
"Potassium," he'd say, "is not really *me*.

I actually feel as if I was meant
To be a *different* element;
Only then might I be able
(Energetically) to be stable."

Well, we all hoped he was wrong, of course,
So we took him to Dr. Strong Force,
Who very carefully analyzed whether
My brother's nucleus could hold together.

He proceeded to write a detailed note on
Every neutron, every proton.
"The reason," said he, "that your brother does mope
Is that he is a different *isotope*.

But the chatter I'll skip to tell you the fact of
The matter: Your brother's radioactive.
Someday he'll undergo beta decay,
But that could be billions of years away."

Said he, with decisive authority,
"Your brother will turn into calcium-40.
A neutron will 'change' to a proton; expect
A particle beta to quickly eject."

We brought him back home, feeling all strange.
We never knew when he would "change"—
While in the solvent swimming pool?
While wooing chlorines at our school?

Year by year, how time did fly.
Then one day, a calcium came on by.
"Oh, please don't mistake me to be another,
For I'm none other than your dear brother!"

I have to admit that I felt really scared.
I knew this would happen, still wasn't prepared.
My brother—with whom I'd the closest of ties—
Was someone I now could not recognize!

"But don't be too tied to my looks or my label.
What's most important is that I am stable!"
I smiled, as I knew things would soon be the same,
'Cause I love my brother, no matter his name.

THE NUCLEAR FAMILY

Little old me really wanted a guy,
So a dating service I thought I'd try.
But I couldn't imagine what I was creating
When I called "Radioactive Dating."

The clearly blatant warning label
Read, "These atoms are unstable!"
But desperate old me really needed a mate,
And so with an atom I went on a date.

And later I married that atom of boron.
I now see I wound up in love with a moron,
'Cause just when I thought he was planning to stay,
The physical laws simply whisked him away.

He underwent a beta decay,
And energy rapidly barreled my way.
I was warned but never cared
That "E equals mc^2."

So now I am feeling incredibly dumb,
'Cause a atom of *carbon* he now has become,
And the lovely husband that I once knew
Has changed his name for good, 'tis true.

And my son, who is troubled, is now on a mission
To rapidly undergo nuclear fission.
Against all the wishes of nurturing Moms,
He wants to go out and to make some bombs.

He hangs out with criminal kids like uranium,
Who put violent thoughts right into his cranium.
With a husband who's changed and son who's deranged,
It's easy to see why *I* feel estranged.

And my daughter's life is also not great.
She's truly obsessed with losing weight.
"Mom, *alpha* decay is the best way to lose,
'Cause now every time I lose *four* a.m.u.'s!"

I ask, "Why me?" but get no answer;
This radiation could lead to cancer.
You think a "nuclear family" works?
You're short by more than a couple of quarks!

DAYS OF OUR HALF-LIVES

My Dearest Love, I'm writing you
To tell you all that I've been through.
I've changed my whole identity,
But loved I can't pretend to be.

When I was uranium-238,
You constantly bugged me to start losing weight.
For five billion years I would hope and I'd pray,
And I finally underwent alpha decay.

Two protons, two neutrons went right out the door,
And now I was thorium-234.
But my nucleus still was unfit for your eyes,
Not positive enough to maintain its big size.

But this time my half-life was not very long,
As my will to decay was now really quite strong.
So it took just a month and not even a millennium
To beta decay into protactinium.

But still you rejected me, right off the bat.
"Protactinium … ? Who's heard of *that*?"
So a beta decay I accomplished once more
To become a uranium-234.

Myself once again, but a new isotope.
You still weren't happy, but I still had hope.
Three alpha decays—'twas hard, but I stayed on
Through thorium again, then radium, then radon.

I thought I would finally satisfy you.
My mass was a vigorous two twenty-two.
But you said, "Although I am pleased with your mass,
I don't want to be with a noble gas!"

So you had a point; I wasn't reactive.
In order to please you, I stayed proactive.
A few days went by, and I found you and said,
"I'd two alpha decays, and today I am lead!"

You shook your head. You weren't too keen
On my latest mass number (two hundred fourteen).
"I've had some bad luck with that mass before,
When an astatine swiftly walked right out the door."

In order to change, I went far away,
But all I could do was just beta decay.
My hopes and my dreams now began to go under,
'Cause beta decays don't change a mass number.

To bismuth, polonium, but hope I had beckoned.
My half-life was one-six-four microseconds.
I finally alpha decayed, and then,
I was lead with a pride-worthy mass of two ten.

I've got to admit I was getting quite tired.
My patience with you now had nearly expired,
As you were more picky than any I'd dated,
And much of my energy had been liberated.

And you *still* weren't happy, but you had a fix.
"I fancy the number 'two-oh-six.'"
So I waited for years until the day
That began with another beta decay,

And then one more still, and then 'twas the end,
When I alpha-ed to lead-206, my friend.
To change any further, I wouldn't be able,
Was no longer active, but happily stable.

It took me some billions of years to do,
But look how I changed, and all just for you!
And what did you say? "You've gotten so old.
I'd like to be with a young lass of gold."

Well, I give up. We're through, my pumpkin!
Shouldn't all my efforts now count for something?
Well, you won't be able to rule me no more;
I'm leaving you—not for one atom, but *four*!

That's right—while you were out blithely diffusing,
I met some nice chlorines whom I found amusing.
And we're gonna form $PbCl_4$,
And you won't be hearing from me anymore.

See over the years, I've grown rather wise.
I've learned that love's about compromise.
You still have half of your half-life to live,
So now you go out there—it's *your* turn to give!

THE RADIOACTIVE DATING GAME

I used to sleep 'til my electrons would drool
At P-32 *element*-ary school.
The things we were taught were just totally boring.
A *mole* of us atoms would always be snoring.

But one thing I learned there I've kept to this day:
"Soon, my students, you'll beta decay
To become more mature and to capture the label
Of 'S-32,' and then you'll be stable.

And when that time comes you'll want celebration,
For you will be ready for graduation.
So look around now, and do count every peer.
Today there's a mole, but you'll soon disappear."

So I watched as my friends all around me decayed,
And I felt left behind, slow and dismayed,
Abandoned by those who were thought to be deft.
In two weeks there was only *half* us of left.

And meanwhile, I'd hoped to impress my young lass
But was now in the bottom half of the class.
I spent every school day conspiring to court her
And remained, two weeks more, in the leftover quarter.

But two weeks more yet, and I still hadn't parted.
Now six weeks had passed since this challenge first started.
Oh, what was my problem? I kept losing faith.
The fraction of atoms now left was one eighth.

Another two weeks and my hope for love waned.
One *sixteenth* of a mole of us atoms remained.
My lass had now probably found a new mate.
By the time I escaped here, it would be too late.

So I studied the past eight weeks with great courage,
And quickly, a pattern began to emerge.
I realized, from evidence existential,
That the decay of the class size was *exponential*!

See, every time two weeks came and then went,
The class size diminished by fifty percent.
'Twas one mole times e to the minus kt,
Where k was *ln* 2 over 2 weeks, you see.

And t was the time, in units of weeks,
Since the teacher that lesson one fine day did speak.
The equation did serve as a useful tool
To predict, at time t, the number in school.

But then it all happened—I had my decay!
I was "S-32," and I liked it that way.
Success carried with it the sweetest aroma,
My electrons excited as I got my diploma,

Which oddly contained a most curious addition.
It read, "You've earned honors and recognition
For thoroughly doing the mental athletics
To uncover the inherent first-order kinetics."

When my love saw this honor she screamed out loud,
"Oh, my brilliant darling, of you I'm so proud!
Your wits have won over my heart in a snap.
Let's let our orbitals overlap!"

So the one thing I learned from that school, 'twas the worst—
That it need not matter who finishes first.
Everyone blossoms at different rates.
(But those who learn chemistry will always get dates.)

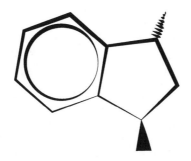

XII. Organic Chemistry

THE ENANTIOMER IN THE MIRROR

Danny heard teasing and taunting galore
From classmates like methane and CCl_4.
He felt unattractive by everyday metrics,
'Cause he was entirely asymmetric.

His carbon was bonded to four different things;
He longed for the beauty of benzene rings,
Was jealous of carbon tetrachlorides,
Whose symmetry served as a sight for sore eyes.

He tried to convince all the lasses that he
Was okay despite the asymmetry.
But all of the ladies would keep on refusing.
"Your four different groups are just *way* too confusing!"

Then one day, ol' Danny discovered an ad:
"Is your own asymmetry making you mad?
Then take on a pose, and model a stance;
Join others like you, and learn how to dance!"

So Danny hoped for some self-esteem
From the "Asymmetric Dancing Team."
His heart did the teammates warm a lot,
For they shared his chemical formula.

The dancing instructor so tastefully
Diffused all about so gracefully.
"Now let us dance," she brightly yelled,
"All $CH(CH_3)FCl$'s!"

Danny was absolutely excited,
For never he'd similar molecules sighted.
So now he could not keep his eagerness in.
"I'm finally going to really fit in!"

But then, as the dancers composed a straight line,
The massive formation was no longer fine.
"Hey Danny, what's wrong with your orientation?
This line must be perfect with no deviation!"

So Danny then flipped all around and turned,
But soon, he became extremely concerned.
Although he'd the very same groups attached,
The others' appearance he just couldn't match.

But their dancing partners soon arrived—
Lasses with thermodynamic drive.
And Danny was different, so others they chose
Out of fear of crushing his 'lectronic toes.

So Danny escaped to the dressing room
And stared at himself with a hopeless gloom.
But then he discovered that in the mirror,
He looked as the others in life did appear.

Facing the glass in a mental scrimmage,
He realized that he was their mirror image.
He couldn't possibly match their pose—
Reflections cannot be superimposed.

At least, this is true for compounds like he,
Who lack a plane of symmetry.
Like left and right hands, they cannot align,
And to the wrong troupe had he been assigned.

His hopes now began a downward spiral.
"So this is what it's like to be chiral.
Not only am I rather unattractive,
With these chiral lasses I'm just not reactive!"

But then, a disconsolate woman came in
With sobbing electrons displaying chagrin.
"This dance class is full of such constant frustration,
'Cause I do not match their configuration!"

Danny looked up, "Hey, neither do I!
So you're not alone. Please don't cry!"
He realized he really did fancy her,
But was she the right *enantiomer*?

"As we both apparently happen to be
The 'wrong' stereochemistry,
Then maybe ..." he paused in mid-advance,
"For each other, we're right. *Shall we dance*?"

And together the chiral compounds united,
Their asymmetric carbons excited.
The couple, they looked in the mirror to see
Their enantiomers smiling ecstatically.

ALL MY CARBONS

Peter Propene was aching with tension;
He *really* wanted to get an extension,
For every day his ego would shatter
Whenever he heard that "size does matter."

So he called BH_3 and then started to woo
To the romantic music of H_2O_2.
When hydroxide attacked, he no longer was small,
'Cause the former alkene became alcohol.

But old Peter Propanol still wanted more,
So he stalked HBr and went right to its door.
His "O" stole the proton he had his keen eye on,
With naught left behind but a bare bromide ion.

And the mad Betty Bromide wanted it back,
So she planned an elaborate backside attack.
But instead of retrieving the H^+ she knew,
She kicked water off in a wild S_N2!

But Peter and Betty, they soon fell in love
With a bond that was strong, like an "orgo" lab glove.
They reckoned that nothing could come in between them,
Not knowing that "Mg" and ether had seen them.

So just when their lives took a happy new start,
Along came magnesium who pushed them apart.
And Peter cried out, "Oh, my life is so hard.
I just wanted love; instead, I'm a Grignard!"

But his pain, although strong, would simply not last,
For a Grignard reacts quite remarkably fast.
So he ran out to find a nice lean foxy bride
And wound up enamored of ethylene oxide.

Was Peter the Grignard just looking for sex
When his desp'rate electrons attacked the vertex
Of this two-carbon triangle? He thought not,
Because married they stayed, and he loved her a lot.

And her oxygen now was quite free from the pain,
As the marriage released all her "ring strain."
And Peter was happy, 'cause now he was large,
And some acid took care of their negative charge.

But life was now taking them out for a ride,
'Cause along came the infamous tosyl chloride.
And what *was* a bad leaving group now was the best,
As the stable –OH had become –OTs.

Old Pete was upset at this shameless invasion.
A hit man he hired for its elimination.
The hit man had spent many years with a trainer
To learn how to attack from antiperiplanar.

Tert-butoxide's the name of this hit man of fame.
To him, tosyl's life was no more than a game.
He snuck in at midnight and checked out the scene
And realized his mission: to make 1-pentene.

He went up to a small beta-hydrogen neighbor
And asked him, "Hey man, can you do me a favor?
You're now my apprentice; I'm gonna teach you
How to kick off a tosyl by using E2.

You'll see that it's really a quick, one-step thing.
As soon as I pull at you, you're going to swing
Your electrons toward tosyl to force it away;
Then we'll leave together and hide, okay?"

So that was the plan, and all went just great.
The tosyl group left; the hit men escaped.
They felt so much pride; they felt that this mission
Was easier still than a *cycloaddition*.

At the end of the day, good ol' Peter was free,
Was no longer in love but still happy, you see.
Although his aloneness he could not ignore,
At least he was longer by two carbons more.

XIII. Quantum Mechanics and Orbitals

GUIDING LIGHT

Hello there, good Doctor. My name is Leon,
Emitted last night by an atom of neon.
Because I am light, I've got reason to rave,
For I can exist as both particle and wave.

Before, I was dating a pretty blue photon,
But we weren't on the same wavelength, and so I rode on
At three hundred million meters per second,
A much faster speed than you ever could reckon.

Whenever my friends ask me, "Leon, what's *nu* ?"
I say, "Five times ten to the fourteenth hertz … and you?"
But I envy the violets, the UV, the blue,
'Cause they have more energy, more than I do.

But I'll get to the point, Doc, 'cause you're in a rush.
My problem is simple: I've got a huge crush
On an atom whose life I could make a lot brighter
If only she'd let in my light to excite her!

Her name is hydrogen—the simplest of matter.
If I try her, at best, I Rayleigh scatter.
What's that you say … it's useless to panic?
It all has to do with … *quantum mechanics* ?

Her energy values, you say, are *quantized*,
Like rungs with spacings specifically sized,
And only those photons whose "E's" match a spacing
Can get her electronic heart to start racing?

So if I were red, or purple, or cyan,
Then I could get her electron to buy in.
But pitiful me, my color is yellow.
I guess I can never be named as her fellow.

And even those colors would pleasure her not,
Unless she already was awfully hot.
At three hundred Kelvin she's in her ground state,
And only UV light can make her upgrade.

I *hate* "Quantum M!" It limits my hopes
To a handful of mates and some isotopes.
Now what's that you say? It's really quite cool,
An analytically useful tool?

As each type of atom has unique spaces,
You shoot it with light, see which *nu*'s it erases.
So all atom types have their own special spectrums.[*]
Whenever they're present, you know to expect 'em.

So look at the light from a distant star,
And you know what atoms it has from afar!
Wow—I guess that my case has been lost;
The benefits clearly outweigh all the costs.

I'll stay *yellow* and mellow, *blue* and teary,
Green with envy, *sighin'* and weary.
Good-bye, my Doc. I think we agree on
The fact that my "E" is still perfect for neon.

For all of you humans out there, think twice
Before you complain that your prospects aren't nice;
Although it seems challenging, finding your mate,
At least *you* don't need physics to resonate!

˙The proper term is "spectra" and not "spectrums."

THE PROTON PERSONALS

"Single, positive, elegant proton
Seeks young and ambitious electron to dote on.
You must be attractive (by Coulomb's Law);
Your wave function cannot possess any flaws.

No baggage or idiosyncrasies, please,
And mass under ten to the minus thirty 'kg's.'
I promise that during our very first date,
I'll give you your choice of an eigenstate.

I'm happy to start out by giving you space.
Just pick out an orbital matching your pace:
The '1s,' if you're certain that I am 'the one,'
The '5f,' *e.g.*, if you're just here for fun.

114

Of all of the orbitals I've got for you,
I thought I would try to describe just a few:
'1s' is the best of the lot, you see;
You get to be *really* close to me!

It may be compact as an eigenstate
But a prime location as real estate,
And if this spherical form you choose,
Then half an atomic unit you'd lose.

But if you are looking for something more spacious,
Then as your good host, I'd be most gracious:
Choose '2s' or any of three '2p's.'
All four are the same in their energies.

So your value of n would be equal to 2.
Your density a node would pass right through.
I hope you'd be fine with that; as you know,
There'd be some locations you could not go.

And whether you opt for the 's' or a 'p'
Will depend on the shape that you hunger to be.
An 's' has a spherical symmetry;
A 'p' looks a tad like a dumbbell, you see.

But if you are not at a point to commit,
Then the '$n=3$' group might be your best fit.
They're higher in 'E' and more spaciously sized,
So you could spread out and delocalize.

There's one '3s' and three '3p's,'
Or you can select from the five '3d's.'
But living as one of these curious abodes
Will mean that you'll need to allow for two nodes.

I honestly hesitate, telling you more,
As I fear that it must be a tedious bore.
And if you decide on a higher state,
We might as well not even bother to date!

But I know for sure (it's a fact, you see)
That once you start getting a sense of me,
You'll rapidly fall in love, say 'yes,
I'll settle right down and choose the '1s'!'

See, thanks to our physics, I now am able
To know what can make you happy and stable.
Once you experience my nuclear caress,
You will not escape from the lovely '1s.'

And because the temperature's so very low,
I know that you'll love me and never will go.
Many say love's got strange dynamics—
Not so if you know your quantum mechanics!"

ME, MYSELF, AND PSI

The other morning, I went to the shrink,
My electronic self sitting right on the brink
Of crossing the fuzzy sanity border
With "undefined energy value disorder."

All of my friends were in their right minds:
"I'm '1s'!" "I'm '2p'!" "My energy's defined!"
But I was not meant to partake in their fate,
Not meant to exist in an *eigenstate*.

116

My wave function summed the "1s" and "2p"
With weights of root 2 and root 1 over 3.
I'd no energetic identification
(By the Copenhagen Interpretation).

The protons I'd dated, from near or far,
Had wanted electrons who "know who they are."
My current proton kept yelling at me
To go get myself an identity.

So I went to the doctor, feeling quite hesitant,
Afraid of experiencing an energy measurement.
But he said, "With just a few finger snaps,
Like magic, your wave function's gonna collapse!"

"But doctor, who will I be when you're through?"
He shrugged and stated, "I wish I knew,
But I certainly do possess the ability
To know every outcome's probability.

See, you are a normalized superposition,
So finding your odds is a simple mission.
I can't know the outcome—I'm *not* omniscient,
But the chances are squares of your coefficients.

So your chance is two thirds of becoming '1s'
And half that for '2p'—quite a bit less.
So sit back, relax, and prepare for the measure.
Here are some books for your reading pleasure."

The next thing I knew the procedure was finished,
And one of my personas had been extinguished.
I found out my energy rather soon, "It's
Negative one-eighth atomic units."

I went to my proton and told her with glee,
"I finally have a defined energy!
I could have been one of two things, you see,
But now I'm one hundred percent '2p'!"

I thought she'd be happy. Instead, she frowned.
As soon as she spoke did my world turn around.
"I dated you just 'cause I knew you were able
With nonzero chance to be *maximally* stable.

I wanted to be with a ground state '1s,'
And your chance of that was two thirds. I guess
That quantum mechanics had forced me to gamble,
But playing the odds has now turned things to shambles."

I told her if only she'd patiently wait
Some nanoseconds, I'd radiate
To '1s,' but this plan we didn't agree on;
To atoms, that timescale is more like an eon.

So thus she sealed off all her dreams with a kiss.
I now know why ignorance often is bliss.
Preceding a measurement, dreams can still be,
But Schrödinger's cat is now laughing at me!

AN N_2 LOVE

Oh, how do you even describe the connection
Between twin sisters? It's really perfection!
Per mole, our bond, which love did make
Takes nearly a million joules to break.

'Tis true that my heart would be crippled, on
Its way out, without our triple bond—
To think of how life would be horrible
With lonely *atomic* orbitals!

Our "1s" electrons are frankly too near
To our nuclei to aid in our friendship so dear,
So the reason that we as two sisters do jell
Is entirely due to our valence shell.

Our "2s's" can link up in two different ways,
Combining both *in* and *out* of phase,
So they are not really worth cheering about,
With the bond and the antibond canceling out.

So allow me to tell you, if you all please,
How the strength of our bond comes from our "2p's."
Though some of the details are still an enigma,
Our "2p_z's" form a bonding sigma.

That this type of bond forms is really quite fortunate;
It's strong, and it forms 'long the bonding coordinate.
But though it's amazing in all of its glory,
This linkage accounts for just part of the story.

Our other "p" orbitals, in particular,
Which to the sigma are perpendicular,
Exhibit true sisterly love and affection
Because of their orbital intersection.

This cozy connection of which I'm fond
Results in *two* terrific pi bonds,
Where orbitals "p" overlap from the side
For *molecular* orbitals worthy of pride.

Each pi bond has two tracts of sisterly love,
Below the bond axis as well as above.
Our "p_x's," they bond, as do our "p_y's,"
And this plays a role in our very close ties.

Because our bond is so hard to sever,
I hope we'll remain as N_2 forever.
But I've heard of stories of forced separation,
The process of *nitrogen fixation* !

And then there's the Haber process, of course,
Producing ammonia by using great force.
When *you* play with chemistry, please have a heart,
And think of the sisters you may force apart.

MOLECULAR MODELS

Today, all our models for you will share
The hottest new styles in molecular wear—
How various particles, tiny and large,
Do stylishly tote their own negative charge.

Diffusing in front down the runway: N_2,
With a look that says, "Watch out, I'm heading toward you!"
Its electronic fabric's a strong triple bond,
The negative cloud so symmetrically donned.

It's bold, but a classic and stable choice.
But now here's a look that will make quite a noise.
It's meant to take people's eyes for a ride
And modeled by hydrogen fluoride.

Notice the risky asymmetry
That says, "*Look*, I act on a whim, so free!"
Electrons are draped 'round the "F" with great fuss,
While the "H" is a nearly exposed H^+!

In order to pull off this look that you see,
You'd need different "e"-negativities.
The atom withdrawing steals "e's" like a crook;
Without this imbalance, you can't get this look.

Next, turn your eyes to the realization
Of 'lectronic orbital delocalization.
We'd like to believe that we're setting some precedents
By dubbing this sleek new summer look "resonance."

To model this look, we chose benzene, the rebel,
Whose skill at pi-bonding has reached a new level.
Delocalized orbitals make it quite able
To keep its electrons extremely stable.

Look at that perfectly planar ring!
This look might become the latest thing!
If *you* want to copy this look so prized,
It helps to be sp^2 hybridized.

Finally, here comes our friend, an O_2,
Who's wearing a fabric that seems rather new.
This fabric is natural, not synthetic.
We're calling this fashion "paramagnetic."

Look, when we turn on a magnetic field,
It's attracted like magic; to none will it yield!
To fashion this stunt, so many had dared;
To succeed, you must have electrons unpaired.

And O_2's got two such electrons, that's why.
They inhabit the antibonding pi,
Creating this look. If you want it, good luck,
For if you are diamagnetic, you're stuck.

So as you can see, my little dearies,
To capture these styles, I've used different theories.
Describing electrons and bonds is quite tough,
So often one model just isn't enough.

But I know these fashions will speak on their own,
So if you are watching this, pick up the phone.
We charge by the charge; 'tis a small price to pay,
So get your electronic outfit today!

XIV. Biochemistry and Cellular Biology

ETHANOL, MY CHILDREN!

Edward Ethanol had nothing to fear
As he floated around in that bottle of beer.
The water was cool, and nowhere in town
Were there enzymes to possibly break him down.

His two carbons free, his hydrogens strong,
His dangling –OH was admittedly long.
His life was complete with bliss and elation
(Because of the process of fermentation).

But one day the cap on his world had been popped;
The bottle was almost entirely flip-flopped.
He fell down a passage and started to shiver
And found himself traveling straight toward a liver!

A sign read, "All alcohols, please be advised;
Beyond this point you'll be oxidized."
And soon he was looking right into the face
Of alcohol dehydrogenase.

The enzyme cut off his hydrogen pair,
And Ed retorted, "Hey, that's not fair.
You may have produced acetaldehyde,
But if I were you, I would run off and hide!"

A second enzyme, to speed up the rate
Of turning ol' Edward to acetate,
Was rapidly forced to display a sign:
"This enzyme is busy. Please wait in line."

It turned out the system had too many vandals,
More aldehyde than this good enzyme could handle.
So Edward now welcomed some time to be free.
He began to go out on his vengeful spree.

He sailed through the blood and pillaged and plundered.
The helpless victims then hopelessly wondered
Why cells all around them started to die,
While overworked enzymes continued to cry.

At the end of the day, Ed declared he had won.
He wreaked tons of havoc and even had fun.
And all of this wreckage was simply to say,
"Cut off my hydrogens, and you'll have to pay!"

PC ARRGHH!

Gary Guanine was totally fine.
He had a great spot in the helical line,
Was paired with a "C" and sat next to an "A,"
So glad to be part of a DNA.

And he was in love with his cytosine mate.
Through hydrogen bonds they'd communicate.
"Oh, aren't you grateful for all of this?
We're happy and safe from hydrolysis!"

He used to be part of an RNA strand,
A second'ry structure partner in hand,
But the strand had degraded away, to his rage,
Because of the ribose's 2-prime –OH.

But now he was happy and finally able
To live in a place that he found rather stable.
But Gary knew only a part of the story;
He actually lived in a *laboratory* !

And one day he found himself in a pipette.
When elsewhere he landed, he started to fret.
He was mixed in with friends in a rapid deluge
And whirled 'round and 'round in a centrifuge.

In truth, the worst had not even arrived,
'Cause of his sweet love would he soon be deprived.
It quickly got hot; he was facing great danger.
The heat caused the DNA pairs to denature!

And just as he called to his partner to stay,
The heat moved the single strands every which way.
"Come back, my love, my shining 'C,'
Don't leave in the name of *entropy*!"

He noticed the temperature, now getting lower;
The rate of his movements was quite a bit slower.
But his love would stay lost 'til the end of all time,
'Cause along came a monstrous *enzyme*.

And soon, ol' Gary was face to face
With a DNA *polymerase*.
"And why are you here?" asked Gary, quite scared.
"I'm here to ensure that you end up base-paired.

See, I'm the best matchmaker you might have seen,
And I'd like you to meet your new cytosine!"
But the new mate was obviously not the same,
Not nearly as nice as his earlier dame.

Her pyrimidine ring, it kind of sagged.
Her backbone was radioactively tagged.
"Why from my love am I forced to be far?"
She answered, "Oh Gary, it's PCR!

Polymerase Chain Reaction's a way
To amplify portions of DNA.
They heat a piece up and break it in two,
Then cool it so primers will bind it like glue.

The polymerase then adds the matching bases
By putting them into their proper spaces.
By repeating this process, I'm happy to say,
You get countless copies of DNA."

But Gary was sad from this shady deal.
"Don't humans consider how *we* might feel?
If *we* went and broke up each *human*'s marriage,
They'd *then* understand why we all must disparage!"

So Gary thus cycled through melting, annealing,
So many a time, with this woeful feeling.
Just hope, as a human, it's actually true
That no one is doing experiments on *you* !

FINDING AMINO

Today I'd like to recount the tale
Of a search that took place on a molecular scale.
A protein, so very determined was it
To find its missing N-terminus bit.

A bouncing baby glycine was he,
Who disappeared so mysteriously.
Did hydrolysis cause this situation,
Or was it by Edman degradation?

So on the morn of that fateful day
Did thus his former neighbor say,
"Together we all must tightly band,
Each alpha helix and beta strand.

We'll search throughout this aqueous medium,
Too vested to notice the mind-numbing tedium.
The solvent is really a dangerous place
For tiny amino acids to face!"

These words were said by a lysine so strong,
Its side chain four carbons, one nitrogen long,
Which, within the protein, had found a nice mate,
Now forming a salt bridge with glutamate.

Their love now disrupted and put to the test,
As they had to spread out and search hard in this quest,
For the slightly salty solvent was vast
And filled with proteins floating past.

As one came by, they asked with a sigh,
"Have you seen a glycine diffusing on by?"
"Oh, *many* … you're playing a losing game,
'Cause, really, all glycines, they *do* look the same.

And also, I fear that unless he's disguised,
He'll probably be *metabolized*,
As I've heard stories, some scary things,
Like glycines *forced* to make *purine rings*.

But luckily, such things are hard to believe, though,
As currently, we are not *in vivo*.
So life is less dangerous; that's how it works.
See, living in test tubes indeed has its perks!"

The passerby left, leaving lonely, lost hearts.
Like flotsam and jetsam, organic spare parts,
Would glycines swim by, from far and wide,
But to cries of "Amino!" no one replied.

"Amino! Amino!" the water shook.
They tumbled through phase space, continued to look.
Then one day a protein, not one that they knew,
Thus bound to them, blocking off half of their view.

129

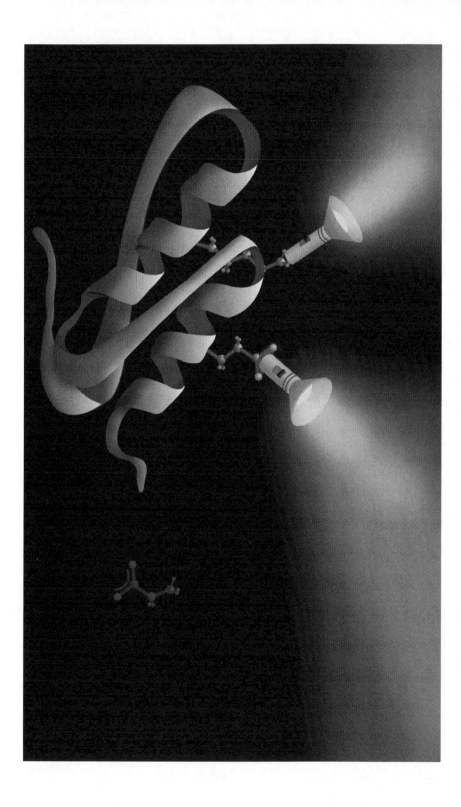

"Hey what are you doing? Get out of here!
We're looking for someone—you'll just interfere!"
"But look," said a voice, "We perfectly match—
I fit in your hydrophobic patch!"

That voice … it sounded so calm and sublime,
What Amino's voice could evolve to in time.
But did he remember them? *Did he know* ?
The lysine so nervously whispered, "*Amino* ?"

Within the peptide backbone they sighted
Amino's alpha carbon excited.
"My beautiful family, oh, so dear!
Alas, I've been safe; you've nothing to fear."

A sigh of relief and the complex relaxed;
Amino survived since the day he'd been axed.
Though they missed the closeness covalence did bring,
Their *van der Waals* bonds were the next best thing.

Their little Amino had mentally grown.
With courage, he showed he could live on his own.
But still it's so tough, as humans we know,
For loved ones to slowly but surely let go.

CRAZY GLU

Once upon a protein's surface,
A glutamate lived who was constantly nervous.
As much as his neighbors would try to appease,
This jittery side chain could not be at ease.

He'd wiggle and stretch, he'd bend and twist,
Abandoning waters he'd recently kissed.
Because his wobbling never ceased,
His protein's entropy stayed increased.

"Hey Glu, calm down. Be still, I say!"
Said a lysine ten angstroms away one day.
"I wish I were close so you'd see my cute face;
Your attraction to me would then keep you in place!"

But he thrashed like a reed who would dance with the breeze,
And soon all his neighbors grew very displeased.
"Ol' Crazy Glu keeps on making a fuss;
He probably thinks he's all *better* than us!"

That night, as they slept in their aqueous stream,
The lysine, she had a most curious dream.
She dreamt how this protein would actually fare
If Crazy Glu were no longer there.

Instead, in his place was an aspartate
Who stayed rather still, in just one state.
Its chain slightly shorter, it could not explore
As much as ol' Crazy could notice before.

The lysine continued to vividly dream
Of a beautiful protein—so real did she seem—

Approaching the lysine and all of her friends.
"My journey has almost arrived at its end.

I'm part of a signaling pathway, you see.
You're next, and as such, you must bind to me
In order to carry the message intact
And deliver the proper, desired impact.

The cell of which we're all a part
Required this specialized pathway to start.
To deliver the output we all must try,
'Cause otherwise (sniff), this cell might die!

So where is that side chain, the one who is deft,
Who flexes to fit in this binding cleft?"
But sadly, there wasn't a single soul who
Could now bind, without Crazy Glu.

The "Asp," he just couldn't stretch quite enough
To bind her at all; 'twas just too tough.
To now interact was the partner averse,
Their "delta G," 'twas now *kcals* worse.*

As she had now dreamt all that she could take,
The lysine found herself awake.
She quickly turned and brought to view,
With great relief, ol' Crazy Glu.

"Keep movin' 'round in overdrive;
You may help keep this cell alive!"
And if *your* peers do norms defy,
Don't chide them first, but find out *why*.

*(per mole)

133

A BEAUTIFUL AMIDE

My favorite phenylalanine,
The prettiest, wittiest gal I'd seen,
Of whom I was always incredibly fond—
We shared such a beautiful peptide bond.

My nitrogen, 'twould entertain her,
Because its groups were fairly planar;
Her carbonyl would oft confide
In me; we made a *great* amide!

One day, the protein of which we were part,
Diffusing naively, without a heart,
It landed itself in the solvent sea
And bound to its arch-enemy!

This enzyme so unkindly rips in
To our view. "I'm chymotrypsin,
And I will do my *very* best,
This peptide chain to now digest."

My 'lectrons shook in utter fear.
"It's going to break us up, my dear!"
And yet, my voice she did not hear;
By then, its active site did see 'er!

I watched her in vain as she struggled in pain
To release her aromatic chain,
But Chymo' quickly managed to lock it
Within a hydrophobic pocket.

And inside a trap was she quickly enclosed,
A serine's oxygen grossly exposed,
Its proton on a nearby "His"
That did in turn an "Asp" chain kiss.

"We accomplish our mission so proteolytic,
For we are a triad catalytic.
The 'His' and the 'Asp' do all the while
Make me a stronger nucleophile,"

The serine explained, and it quickly went back
To a textbook carbonyl attack,
Controlling the complex and quickly leading it
To tetrahedral intermediate.

The negative charge was stabilized
By backbone "H's." I realized
The hopelessness, and I started to panic—
I'd soon be a victim of thermodynamics!

Indeed, the enzyme, it forced me out,
And all I could manage to do was to shout,
"Good-bye, dear friend!" so sadly spoken,
Our wonderful peptide bond now broken.

I drifted away from that active site,
And as my friend put up a fight,
The last I could see was an H_2O
Attacking the carbon I once did know

And forming again an intermediate,
Another broken bond succeeding it,
But this time the enzyme was forced away
To find yet another ol' victim for prey.

And I had no choice but to make a new start,
My friend and I *nanometers* apart.
To think ... my confidante so prized
By now might be metabolized!

But I'll remain positive, right 'til the end,
Still hoping that she has now found a new friend,
Her side chain so happy with optimal packing
And perfect aromatic stacking.

And as I'm a serine, I picture the case
Where I'd soon play a role in a protease.
I'll admit that it's wrong, but I always will say,
"I hope to take my revenge one day!"

DRUG TEST

All of my atoms were quietly shaking,
So nervously fearing this undertaking.
But *focus*, I had a goal, a vision—
To make it past this first audition.

Millions of compounds like me were competing
That day at a crucial molecular screening.
But all of my fears would I have to shrug
If I wanted to be the "next big drug."

The screening call announcement read,
"We hate HIV, so let's come out ahead!
Do you think you have the optimal face
To inhibit an HIV protease?"

My turn had come! I tumbled with care
To face the protease, and there—
Without even needing to put up a fight—
I diffused right into its active site!

And soon, I saw others were rolling on by
And staring me down with an envious eye.
But they were unable to pry me free.
(I'd bound with quite negative "delta G.")

The protease whimpered incessantly,
"Oh, please, I say! Let go of me!
You've brought all my work to a halt, you know,
By sitting right where all my substrates should go!"

I smiled. Perhaps my dreams would thrive.
See, HIV needed his job to survive,
But I clogged up the site with a fit rather snug;
I could stop the virus, could be ... a drug!

Soon, I heard many people exclaim,
"We've gotten a hit!" and then came the fame.
In weeks that followed, they took many actions
To note my inhibiting interactions.

They found that my shape, it was just about right.
My molecular mass was suitably light,
And the hydrogen bonds that I made were all strong,
So none was electrostatically wrong.

My hydroxyl group was forced to crane,
So yes, there was internal strain,
But overall, I felt quite free
To move about entropically.

But just when I thought that the world I was winning,
I heard, "Beware—this is just the beginning.
There are *many* more tests you will have to complete,
'Cause becoming a drug is no easy feat!

Your binding will need to be highly specific
Before they'll consider you truly terrific.
Your affinity must be extremely persistent,
For HIV mutates, becomes drug-resistant."

And sure enough was I told with a sigh,
My toxicity, it was a little too high.
"We really, truly value your patience,
As we make chemical modifications."

I found myself now in a mental jam.
"You're going to alter who I *am* ?"
Was this worth hopping through painful hoops,
Like losing my favorite functional groups?

But then I remembered the ultimate goal
With tingles throughout my molecular soul:
The millions of people with HIV,
I may prolong their lives, you see.

It wasn't about my fame or my glory
But writing a happier end to *their* story.
And that's when I felt really proud to be
A part of biochemistry.

THE ALPHA PERSONALITY

I have to admit, it's irritating!
I spend all my days and nights a-waiting
And losing my mind, going insane,
Diffusing about this cell membrane.

I have the privileged power to tell
Specific messages to my cell.
But wasted away are my skills every day,
'Cause here in this cell, there is *nothing* to say!

I guess it must be isolated,
Receptors rarely activated,
And life becomes a boring scene
For me, inside a G-protein.

But what does this G-alpha protein old-timer
Expect from his duty as part of a trimer?
Well, though I think they're nice, I am a
Little sick of "beta-gamma."

But now the world will all be mine;
I bribed a "Mr. Cytokine"—
He'll bind receptors at the surface,
And then I will finally serve my purpose.

Oh—here he is! What transformations
Of bound receptors' conformations,
Including the one to which I am stuck.
My 'lectrons are dancing, enjoying their luck!

A GTP molecule beckons to me,
And voilà, like that, I'm finally free
And able to relish in what I do well,
To tell all my news to the rest of the cell.

There's limited time, 'cause I need to race
Against my very own GTPase.
What separates boredom from messenger's bliss
Is GTP hydrolysis!

I move all around with breath so bated.
"Must bind to a target and activate it!
Must find all the folks downstream and command 'em.
Too bad that my motion is totally random."

I hope and I pray for productive collisions,
But sadly, my hopes don't achieve their fruition.
For long before any targets I see,
My clock is demoted to GDP!

Morosely, I realize I might not advance
This cellular message because of sheer chance.
How painful it is for me to see
The other G-alphas succeed and not me!

But then do I hear an attractive voice,
Abounding with glee and molecular poise—
New isoforms of beta-gamma.
"Hello," she says, "My name's Suzanna."

The loveliest partner that I could find
Is waiting right here for me to bind!
So though as a messenger I lacked success,
In this trimer I'll find true happiness!

OH, MINE, CYTOKINE

She diffused all around in a random walk
With only electrostatics to stalk,
Bombarded by overly zealous water
And hoping her one true love could spot her.

As time went by, she'd fix her domains,
Perfecting the strands and the loops with great pains—
A tyrosine here, a salt bridge there,
All perfectly poised to attract a stare.

But hours would pass with no love found,
With ev'ry side chain left unbound;
A wave of panic surrounded her core.
"My thermodynamic life is a bore!"

But just when her hopes were all falling apart, her
Path led her right to a binding partner:
Desp'rate, too, and willing to 'ccept her,
A massive *cytokine receptor*.

And toward his protein arms she flew.
Said he, "I've waited forever for you!"
And contact was made, conformations were changed,
Through forces so strong were they both rearranged.

At first she was only the slightest bit nervous
To jauntily cruise on the cellular surface;
Her partner's affairs, she just couldn't tell
With whom he was binding *inside* of the cell.

But still he was bound to her, this was a fact,
Their charges near perfect so they could attract.
He never considered to call it quits
Amidst the clathrin-coated pits.

They soon were surrounded, within a new home,
Internalized into an endosome.
Entrapped in the cell, such treatment unethical,
And forced to remain there within a vesicle.

She began to get scared, was unsure of her fate,
Took comfort in clinging along to her mate.
But something was changing, hour by hour—
It seemed like the marriage was starting to sour.

She felt like her love was now losing his grip
As the pH around them was starting to slip.
"Oh, no!" she screamed. "Don't go! Just wait—
I'll die if you ever dissociate!"

Their side chains fighting, wildly flailing,
"I'm sorry," he said, amidst her wailing.
"Just trust me, in time will you understand."
With that, he let go of her lysine hand.

She screamed as she struggled to stay afloat,
And a wave of solvent engulfed them both.
She watched him be swept by the inundation,
Toward lysosomal degradation.

And then she knew, this widowed wife:
He'd left her side to save her life.
Had she been bound through all of this,
She'd die from proteolysis!

The next thing she knew, she was quickly set free,
Released from the cell ceremoniously,
Along with some cytokine sisters and brothers:
"Our old mates are gone—let's try and find others!"

Her shocked reaction they noticed and said,
"Indeed, it is true that your lover is dead.
But lucky you are, you escaped just fine,
So savor your life as a cytokine."

And off they all went to discover new mates,
With rapid association rates.
But she simply couldn't forget her past,
Those glutamate arms that she held last.

And while she eventually bound again,
'Twas never the same as way back when …
For though it usually ends with a part,
One's first true love remains in her heart.

Made in the USA
Columbia, SC
12 May 2017